S0-CVE-401

ENZYME INDUCTION, MUTAGEN ACTIVATION AND CARCINOGEN TESTING IN YEAST

Ellis Horwood books in the
BIOLOGICAL SCIENCES
General Editor: Dr ALAN WISEMAN, University of Surrey, Guildford

Series in
BIOCHEMISTRY AND BIOTECHNOLOGY

Series Editor: Dr ALAN WISEMAN, Senior Lecturer in the Division of Biochemistry, University of Surrey, Guildford

Ambrose, E.J.	**The Nature and Origin of the Biological World**
Berkeley, R.C.W., *et al.*	**Microbial Adhesion to Surfaces**
Blackburn, F. & Knapp, J.S.	**Agricultural Microbiology***
Bowen, W.R.	**Membrane Separation Processes***
Bubel, A. & Fitzsimons, C.	**Atlas of Cell Biology***
Butt, W.R.	**Topics in Hormone Chemistry Vol. 1**
Cockill, J.A.	**Clinical Biochemistry**
Dean, A.C.R., *et al.*	**Continuous Culture Vols. 6 & 8**
Dunnill, P., Wiseman, A. & Blakeborough, N.	**Enzymic and Non-enzymic Catalysis**
Eyzaguirre, J.	**Chemical Modification of Enzymes**
Horobin, R.W.	**Chemical and Physical Basis of Biological Staining***
Inoki *et al.*	**Dynamic Aspects of Dental Pulp**
Kennedy, J.F., *et al.*	**Cellulose and its Derivatives**
Kennedy, J.F. & White, C.A.	**Bioactive Carbohydrates**
Kricka, L.J. & Clark, P.M.S.	**Biochemistry of Alcohol and Alcoholism**
Palmer, T.	**Understanding Enzymes 2nd Edition**
Reid, E.	**Methodological Surveys in Biochemistry Vols. 6–11**
Roe, F.J.C.	**Microbiological Standardisation of Laboratory Animals**
Reizer, J. & Peterkofsky, A.	**Sugar Transport and Metabolism in Gram-positive Bacteria***
Sammes, P.G.	**Topics in Antibiotic Chemistry Vols. 1–6**
Scragg, A.	**Biotechnology for Engineers**
Sikyta, B.	**Methods in Industrial Microbiology**
Verrall, M.S.	**Discovery and Isolation of Microbial Products**
Verrall, M.S.	**Separations for Biotechnology***
C. Webb & F. Mavituna	**Process Possibilities for Plant and Animal Cell Cultures**
Wiseman, A.	**Handbook of Enzyme Biotechnology 2nd Edition**
Wiseman, A.	**Topics in Enzyme and Fermentation Biotechnology Vols. 1–10**
Wiseman, A.	**Enzyme Induction, Mutagen Activation and Carcinogen Testing in Yeast**

* *In preparation*

ENZYME INDUCTION, MUTAGEN ACTIVATION AND CARCINOGEN TESTING IN YEAST

Editor:
ALAN WISEMAN Ph.D., F.R.S.C., M.I.Biol.
Senior Lecturer in the Division of Biochemistry
University of Surrey, Guildford

ELLIS HORWOOD LIMITED
Publishers · Chichester

Halsted Press: a division of
JOHN WILEY & SONS
New York · Chichester · Brisbane · Toronto

First published in 1987 by
ELLIS HORWOOD LIMITED
Market Cross House, Cooper Street,
Chichester, West Sussex, PO19 1EB, England
The publisher's colophon is reproduced from James Gillison's drawing of the ancient Market Cross, Chichester.

Distributors:
Australia and New Zealand:
JACARANDA WILEY LIMITED
GPO Box 859, Brisbane, Queensland 4001, Australia
Canada:
JOHN WILEY & SONS CANADA LIMITED
22 Worcester Road, Rexdale, Ontario, Canada
Europe and Africa:
JOHN WILEY & SONS LIMITED
Baffins Lane, Chichester, West Sussex, England
North and South America and the rest of the world:
Halsted Press: a division of
JOHN WILEY & SONS
605 Third Avenue, New York, NY 10158, USA

British Library Cataloguing in Publication Data
Enzyme induction, mutagen activation and carcinogen testing in yeast. —
(Series in biochemistry and biotechnology). —
(Ellis Horwood books in the biological sciences).
1. Chemical mutagenesis — Analysis
2. Mutagenicity testing 3. Yeast
I. Wiseman, Alan II. Series
575.2'92 QH465.C5

Library of Congress Card No. 87–3961

ISBN 0–85312–963–0 (Ellis Horwood Limited)
ISBN 0–470–20856–2 (Halsted Press)

Phototypeset in Times by Ellis Horwood Limited
Printed in Great Britain by R. J. Acford, Chichester

Table of contents

1 Editor's Introduction
Dr Alan Wiseman, Biochemistry Division, Department of Biochemistry, University of Surrey, Guildford
1.1 General introduction: testing for mutagenicity and carcinogenicity in yeast . . . 9
1.2 Introduction to Chapter 2–6 . . . 10
References . . . 10

2 Genetic systems in yeast and their application in drug screening
Professor D. Wilkie and D. C. Collier, Department of Botany and Microbiology, University College, London
2.1 Introduction . . . 12
2.2 Life cycle of *Saccharomyces cerevisiae* . . . 12
2.3 Tetrad analysis . . . 17
2.4 Mitotic recombination . . . 18
2.5 Drug resistance mutations: isolation and genetics . . . 20
2.6 The mitochondrial system . . . 22
2.7 Discussion . . . 39
References . . . 39
Appendix . . . 40

3 Mechanisms of gene induction and repression in *Saccharomyces cerevisiae*
Dr P. W. Piper, Department of Biochemistry, University College, London
Abbreviations . . . 41
3.1 Introduction . . . 41
3.2 The synthesis and maturation of transcripts made by RNA polymerase II in *S. cerevisiae* . . . 42
3.3 Systems for identifying transcriptional control elements . . . 47
3.4 Transcript and promoters for RNA polymerase II . . . 53

3.5 Induction of *S. cerevisiae* genes by positive regulation of transcription 56
3.6 Gene activation and repression in the determination of mating type 610
3.7 Conclusions and future prospects 66
References 68

4 DNA repair and mutagenesis in *Saccharomyces cerevisiae*
Dr A. J. Cooper, Department of Pathology, Stanford Medical School, Stanford University, California, USA and Dr S. L. Kelly, Wolfson Institute of Biotechnology, The University, Sheffield
4.1 Introduction 73
4.2 General concepts of DNA repair and mutagenesis 74
4.3 DNA repair in *Saccharomyces cerevisiae* 77
4.4 Photoreactivation 77
4.5 Radiation-sensitive (rad) mutants: assignment to DNA repair pathways 78
4.6 The *RAD*3 epistasis group involved in excision repair 79
4.7 Genes which are induced by DNA damage 85
4.8 The *RAD*6 pathway involved in mutagenic repair 86
4.9 The *RAD*52 epistasis group involved in recombinational processes 89
4.10 The role of yeast DNA repair processes in chemicaly-induced DNA damage 92
4.11 Aspects of mutagenesis in yeast 95
4.12 Conclusions 104
References 105

5 Yeast cytochrome P-448 enzymes and the activation of mutagens, including carcinogens
Dr D. J. King and Dr A. Wiseman[†], Biochemistry Division, Department of Biochemistry, University of Surrey, Guildford
[†]Present address: Protein Biochemistry Department, Celltech Ltd., 244–250 Bath Road, Slough, Berks, Uk SL1 4DY
5.1 Cytochrome P-488/P-450 enzymes 115
5.2 Cytochromes P-450 in *Saccharomyces cerevisiae* 131
5.3 Techniques for the study of yeast Cytochrome P-450 enzymes . 152
5.4 Conclusions 157
References 157

6 Detection of mutagens in yeast
Dr D. E. Kelly, Wolfson Institute of Biotechnology, The University, Sheffield
6.1 Introduction 168
6.2 Test systems and the genetic endpoints 170

1

Editor's introduction

Dr. Alan Wiseman
Biochemistry Division, Department of Biochemistry, University of Surrey, Guildford

1.1 GENERAL INTRODUCTION: TESTING FOR MUTAGENICITY AND CARCINOGENICITY IN YEAST

Although yeast has been used in the traditional biotechnology industries of brewing and baking for many centuries (Wiseman, 1985, 1977–1985), it has been the advent of recombinant DNA technology during recent years that has stimulated the remarkable present interest in its genetics and synthetic capabilities. Important medical products are being made by 'brewing techniques' following the incorporation and expression in the yeast of foreign genes, some of human DNA sequence, by the techniques of genetic engineering (see review by Kingsman *et al.* 1985, 1987).

In parallel with these very important developments in the manufacture of particular proteins, there is another field of growing interest in yeast where the application of genetics is important. This field is in the study of genotoxicity by a variety of techniques. Most carcinogens are clearly genotoxic chemicals and can cause a variety of gene damage related responses that may be readily detected in the yeast. This book therefore is concerned with the use of yeasts to detect carcinogens by observing the mutagenic effects caused by these carcinogenic chemicals. Along with similar approaches using bacteria, yeast is proving to be useful also in the detection of mutagenic agents, and likely carcinogens therefore, in the environment, in foodstuffs, and in pharmaceutical preparations. Yeast cells are eukaryotic, like mammalian cells, and there may be advantages in their use, relative to the use of the bacterial (prokaryotic) cell. In addition, the *in situ* (closer to the DNA) activation of these mutagens (and carcinogens) that require to be chemically activated (modified) before they can damage DNA in the 'appropriate manner' has been demonstrated inside yeast cells that contain cytochrome P-448 enzymes. It may be important not to have to activate the carcinogens outside the cell by using the usual rat liver cytochrome P-448/P-450 additive, with a preincubation period for activation.

1.2 INTRODUCTION TO CHAPTERS 2–6

An excellent account of mutagenicity testing, in a practical approach, has been published by Venitt & Parry (1984). Details of the practical requirements for using yeasts can be found in the chapter entitled 'The assay of genotoxicity of chemicals using the budding yeast *Saccharomyces cerevisiae*' by Elizabeth M. Parry and James M. Parry, pp. 119–147.

Very considerable background knowledge of the molecular biology and enzymology of yeast will be needed to fully understand the complex requirements for the activation of mutagens (and carcinogens) especially when all the test system is inherent within the yeast cell itself. This then includes the understanding of the yeast cytochromes P-448 enzyme system; its production and mode of action in the yeast cell. This will require a knowledge not only of the genetics of the yeast used, but also of the protein biosynthesis machinery and how it may be switched on by induction, perhaps by the mutagen itself. Thus Chapter 2 by Professor D. Wilkie outlines some distinct genetic features of yeast, and reviews the very important mitochondrial DNA system which is important in its own right for the study of petite mutants arising from the use of particular chemicals to damage mitochondrial DNA. Dr P. Piper (Chapter 3) has written a detailed description of the molecular biology of transcription of genes in yeast, including a discussion of transcriptional promoters for RNA polymerase II. This chapter includes also the mechanisms of induction of yeast genes by positive regulation of transcription. This is of interest also in relation to induction of mutagenesis described by Dr A. J. Cooper and Dr S. L. Kelly (Chapter 4). Their detailed account of DNA repair and mutagenesis in yeast, and other micro-organisms, is of great interest therefore in relation to the practical detection of mutagens in yeast (Dr D. E. Kelly in Chapter 6).

Chapter 5 by Dr. D. J. King and Dr. A. Wiseman discusses the nature and roles of the cytochrome P-448 enzymes required to activate many mutagens (and carcinogens). Yeast cytochromes P-448/P-450 should be the subject of considerable research in the future to substantiate and understand their role in the activation of such mutagens, and putative carcinogens therefore. The acceptability and use of the yeast test system is increasing, and this system may become of major importance because of its eukaryotic microbial advantages. It is of interest in this and other connections that the cost-effectiveness and desirability of short-term tests for carcinogenicity of chemicals has been favourably assessed recently. (Lave & Omenn, 1986).

REFERENCES

Lave, L. B. & Omenn, G. S. (1986) *Nature* **324** 29–34.

Kingsman, S. M., Kingsman, A. J., Dobson, M. J., Mellor, J. & Roberts, N. A. (1985) In: *Biotechnology and genetic engineering reviews* (ed. Russell, G. E.) Volume 3 pp. 377–416 Intercept.

Kingsman, S. M., Kingsman, A. J. & Mellor, J. (1987) *Trends in biotechnology* **5(2)** 53–57.

Venitt, S. & Parry, J. M. (1984) *Mutagenicity testing — a practical approach*, IRL Press, Oxford.

Wiseman, A. (1985) *Handbook of enzyme biotechnology*, 2nd edn; and also *Topics in enzyme & fermentation biotechnology* (1977–1985 Vols 1–10) Ellis Horwood.

2

Genetic systems in yeast and their application in drug screening

Professor D. Wilkie and D. C. Collier
Department of Botany and Microbiology, University College, London

2.1 INTRODUCTION

Genetic analysis is not usually an end in itself nowadays but is frequently used as an adjunct to biochemical investigation. Many of the advances in our knowledge of the genetic material and its function in the control of biological processes have required the production of mutants, while the detection and characterization of mutants *per se* is of importance in assessing mutagenic potential of drugs and chemicals. Also, the study of the mode of action of drugs at the cellular level is greatly facilitated by the isolation of mutants showing resistance to their toxic effects, while in the field of genetic engineering linkage data can be useful in locating recombinant DNA.

In the yeast *Saccharomyces cerevisiae*, genetic analysis is relatively straightforward, based as it is on meiotic segregation with the availability of tetrad analysis as an added bonus. Also, this organism has the unusual feature of being stable both in the haploid and the diploid condition. The latter allows recessive/dominant relationships of genes to be ascertained, putting it on a level with higher eukaryotes, as well as opening the way to genetic analysis *via* mitotic recombination. An outline account of these basic concepts, the techniques used, and the interpretation of segregation data will be given, with the non-specialist particularly in mind. Criteria for distinguishing between nuclear and mitochondrial genetic systems and their interactions will also be listed and described. The application of these systems to the screening of drugs will be demonstrated.

2.2 LIFE CYCLE OF *SACCHARYOMYCES CEREVISIAE*

S. cerevisiae is unicellular and grows vegetatively by budding. Strains exist as haploid cultures, but when cells of strains of opposite mating type come into contact they can fuse in pairs to form diploid zygotes. Control of the mating reaction is vested in two alleles, *a* and α, of a single nuclear gene. Thus, mating will take place between two strains if one carries *a* and the other α, and the zygotes (Fig. 2.1a) will all be heterozygous a/α. If mating mixtures are set up in growth-promoting medium, zygotes will bud off

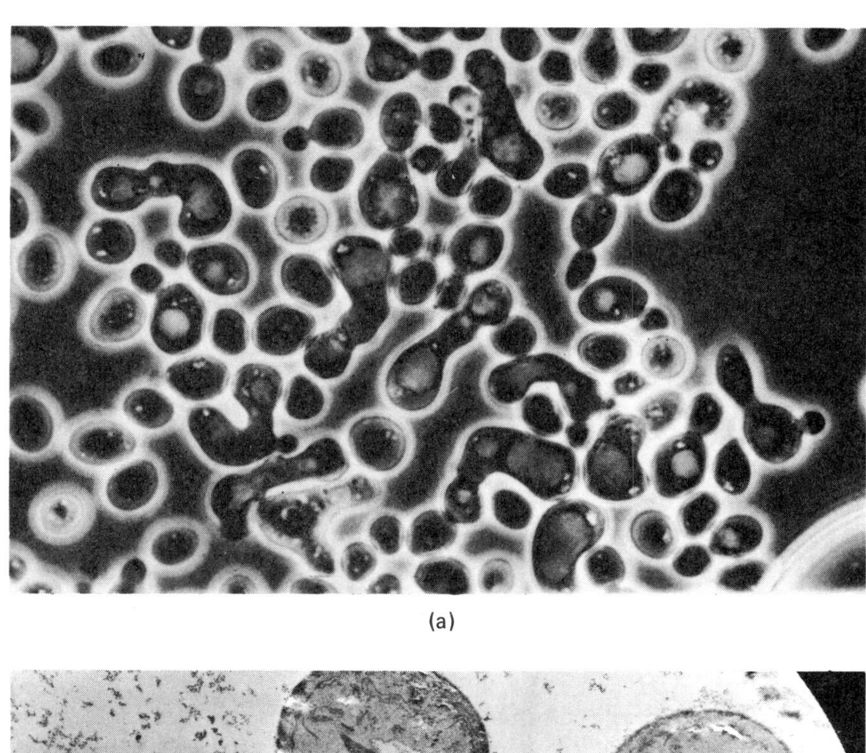

(a)

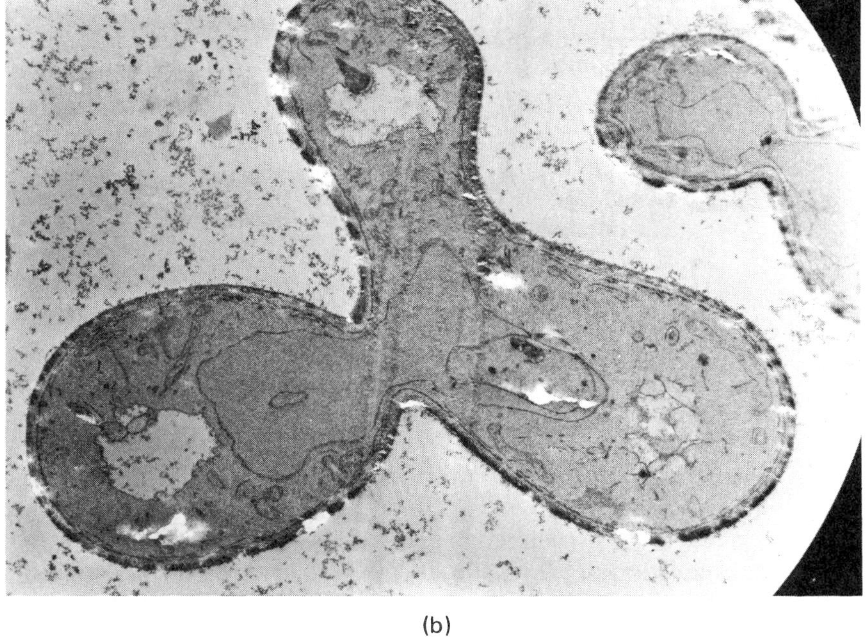

(b)

Fig. 2.1 — (a) Mating mixture of two haploid strains of *Saccharomyces cerevisiae* in the light microscope showing zygotes (4 h. in YED medium). (b) Electron microscope picture of a zygote budding off a daughter cell. Note mitotic nucleus still contained in the nuclear envelope.

diploid daughter cells (Fig. 2.1b) and diploid cultures will be established. The isolation of diploid clones is greatly facilitated by mating strains carrying different auxotrophic markers. Heavy suspensions of each strain are mixed in water and several drops inoculated onto solid minimal medium (for all media, see Appendix). Any colonies that come up after incubation must be diploid since neither parent can grow in unsupplemented medium, discounting reversion to prototrophy which would be comparatively rare. If the biochemical markers are respectively *ade* (adenine deficiency) and *his* (histidine requirement), for example, the genotypes in the cross can be represented as follows:

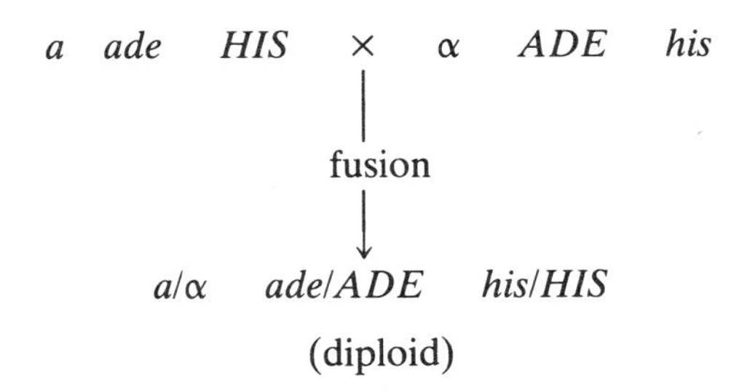

The diploid will be prototrophic for adenine and histidine since the wild type genes (*ADE* and *HIS*) will be dominant.

Strains with biochemical marker genes are available from national collections, but biochemical mutants may be picked up using a technique involving magdala red. This dye is added to a complete agar medium on which cells of a prototrophic strain are plated after treatment with a suitable mutagenic agent (e.g. UV radiation). Auxotrophic colonies tend to take up the dye to a greater extent than prototrophic colonies, and since the dye is not toxic, stained colonies can be isolated for further testing. Details of this procedure have been published (Horn & Wilkie, 1966).

The second stage of the life cycle comprises a meiotic division and a return to the haploid state. Meiosis can be induced in diploid cells by starvation conditions in sporulation medium (SPM). Each diploid cell embarking on the reduction division becomes an ascus, and the mature ascus normally contains a tetrad of ascospores (Fig. 2.2). The first stage in the series of events leading up to tetrad formation is the replication of individual chromosomes resulting in bivalents held together at the centromere. The next stage is pairing up of homologous bivalents with coincidence of base sequences along their lengths. It is at this point that crossing-over can occur between chromatids, the name given to each member of a bivalent. Since these exchange events apparently occur at random, then the further apart two genes are the more frequent will a cross-over occur between them to give a recombinant chromatid. It is thus the frequency of this event that establishes the degree of linkage between two genes. If among the meiotic progeny there were 10% recombinants (crossover classes), the two genes would have a map distance of 10 units between them. The same principle

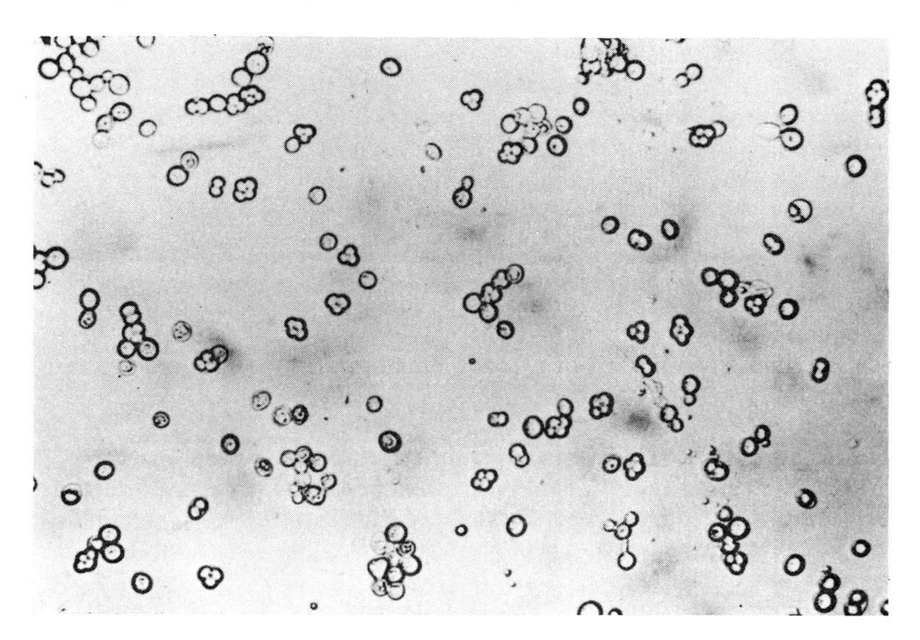

Fig. 2.2 — Diploid cells in sporulating medium showing ascospore tetrads (after 48 h.).

applies in mapping mutant sites within a gene. Independently arisen mutations affecting the same gene will almost certainly occur at different sites within the gene since there are theoretically as many mutant sites as there are bases in DNA making up the gene. Recombination can take place between mutant sites (m) to restore a non-mutant strand during meiosis (see Scheme at top of p. 16).

In the diploid, the combination of m_1 and m_2 is said to be heteroallelic. Furthermore, the heteroallelic diploid will show the mutant phenotype since m_1 and m_2 both have aberrant products; that is, there will be no complementation in the diploid as a rule. Complementation to give wild type diploids would establish the non-allelism of the mutations involved, and this principle is well illustrated in the genetic analysis of the adenine biosynthetic pathway in yeast described by Roman. The synthesis proceeds by a series of steps, each catalysed by a different gene product:

$$\mathrm{A} \xrightarrow{ADE_1} \mathrm{B} \xrightarrow{ADE_2} \mathrm{C} \xrightarrow{ADE_3} \mathrm{D} \xrightarrow{ADE_4} \mathrm{E} \xrightarrow{ADE_5} \text{adenine}$$

Mutation in any one of the *ADE* genes leads to adenine auxotrophy. A number of adenine-requiring mutants were isolated independently in two strains of opposite mating type and systematically analysed in pairwise crosses. When different genes are mutant, complementation (adenine prototrophy) would be seen in resulting diploids. For example, assume the

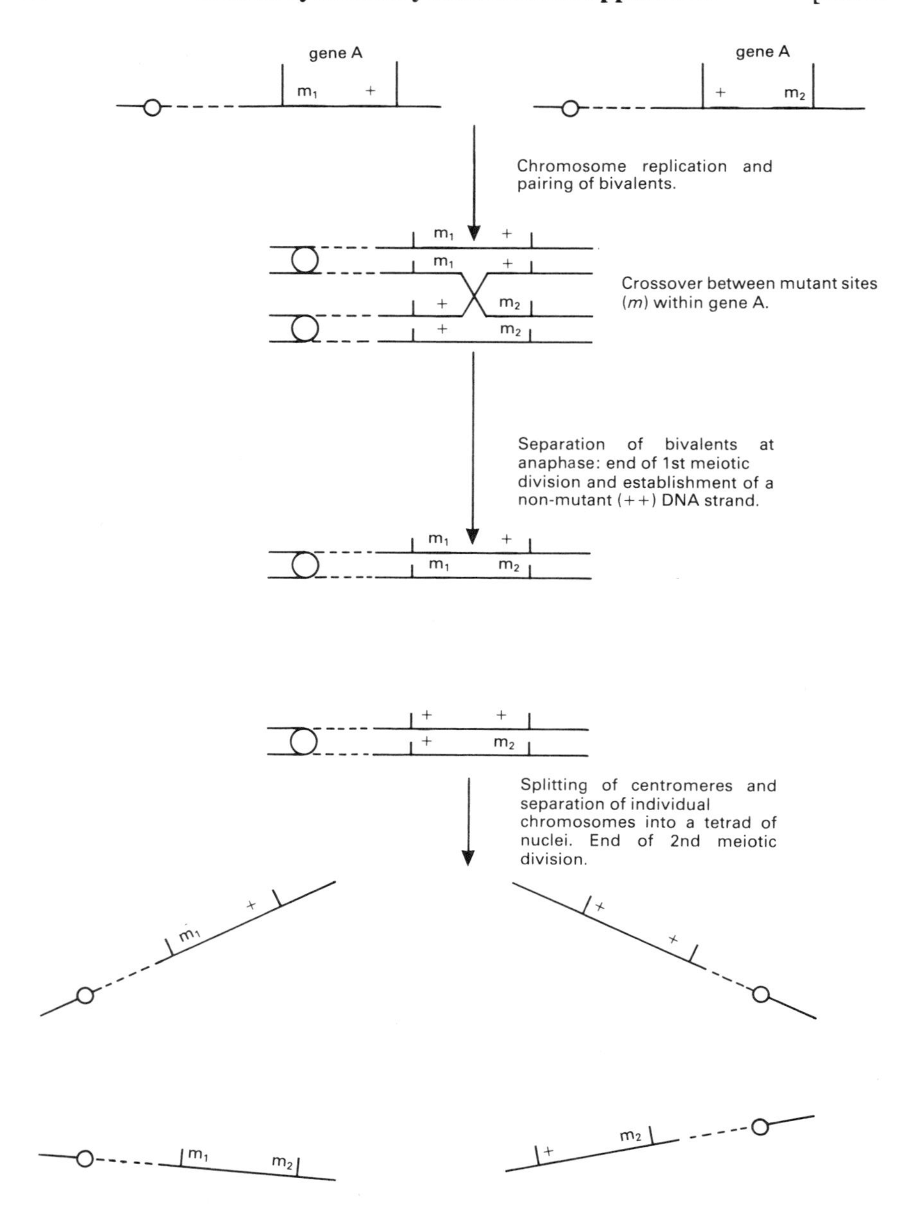

mutation ADE_1 to ade_1 occurred in strain 1 and ADE_3 to $ade3$ in strain 2, then the cross can be depicted as follows:

$$ade_1 \quad \mathrm{ADE_3} \quad \times \quad \mathrm{ADE_1} \quad \mathrm{ade_3}$$

$$\downarrow$$

$$\frac{ade_1}{\mathrm{ADE_1}} \quad \frac{\mathrm{ADE_3}}{\mathrm{ade_3}} \quad \text{diploid}$$

The diploid will be heterozygous for both genes and will have the wild type phenotype, i.e. adenine independence. On the other hand, if there is no complementation, clearly the mutation in each strain has affected the same gene.

2.3 TETRAD ANALYSIS

In the meiotic products of this diploid there are four possible arrangements of the alleles of the two genes, namely, ade_1 ADE_3, ADE_1 ade_3 (these two being parental types), ade_1 ade_3, ADE_1 ADE_3 (these being recombinant types). If the two genes are unlinked, the four classes will occur in the ratio of 1 : 1 : 1 : 1 in a random sample of meiotic products (ascospores), but if they are linked there will be a preponderance of parental types. The recombinant type ADE_1 ADE_3 is, of course, wild type, and its appearance would be predicted on the basis of the complementation in the diploid.

When the four products of individual meioses (tetrads) are microdissected and analysed, three tetrad types will be seen:

1		2		3	
ade_1	ADE_3	ade_1	ADE_3	ade_1	ade_3
ade_1	ADE_3	ade_1	ade_3	ade_1	ade_3
ADE_1	ade_3	ADE_1	ADE_3	ADE_1	ADE_3
ADE_1	ade_3	ADE_1	ade_3	ADE_1	ADE_1

Type 1 is a parental ditype (PD) showing only the two parental arrangements of the genes; type 2 is a tetratype (T) with parental and recombinant arrangements; while type 3 is a non-parental ditype (NPD) showing only recombinant ascospores. If the two genes are unlinked they will re-assort freely so that any one arrangement is as likely as any other arrangement, although it will be apparent that re-assortment to give type 2 tetrad can occur in more than one way. The distribution of tetrad types is very different when two genes are linked: most tetrads will be of type 1, since type 2 can arise only when there is a crossover between the two genes, while type 3 requires a rare double crossover involving all four chromatids. The recombination percent which gives the distance (x) in centimorgans between the two genes can be calculated from the equation

$$x = \frac{100\ (\mathrm{T} + 6\mathrm{NPD})}{2(\mathrm{PD} + \mathrm{NPD} + \mathrm{T})} .$$

2.4 MITOTIC RECOMBINATION

It is possible to obtain recombinant cells in heterozygous diploid yeast without the necessity of going through the meiotic process. The fact that recombinational events can take place during growth of diploid cells was first described fifty years ago by Stern (1936), in the fruit fly *Drosophila.* It was later established that homozygous segregants can arise from heterozygous diploids in general during growth owing to an aberration of the mitotic division. This occurs early in mitosis after chromosome replication when individual homologous chromosomes accidentally pair up and exchange segments as in meiosis before continuing the mitotic process. As illustrated in Fig. 2.3, this can lead to the formation of a homozygous (*a*/*a*) cell in a heterozygous culture (*A*/*a*) depending on the segregation pattern of chromosomes at anaphase. The homozygous *a*/*a* cell and its mitotic progeny would show the mutant phenotype assuming *a* is recessive, while the reciprocal homozygous *A*/*A* cell would be indistinguishable from heterozygous *A*/*a* cells. It will be appreciated that all heterozygous genes on the same chromosome arm as gene *A* but distal to it with respect to the centromere, will also become homozygous following the crossover and segregation as shown in Fig. 2.3. Simultaneous homozygosity is thus indicative of linkage, and the degree of linkage can be assessed. For example, if homozygosity of a murker gene (see below) is almost always accompanied by homozygosity of a gene or genes under investigation, then close linkage is indicated. It is possible, then, to obtain the order of genes and their relative positions on chromosomes using data from mitotic recombination. It is estimated that the aberration occurs about once in 1000 mitoses.

Useful markers in this system include the pigmented, adenine-deficient mutant yeast cells (ade_2) and recessive drug-resistance genes. In the diagram of adenine biosynthesis (above, the ade_2 gene codes for an enzyme which catalyses the step B to C. Blockage of this step in the mutant leads to accumulation of precursor substance B. This forms the substrate for another enzyme system in which the end product is a red pigment, an apparently aimless process which slows growth rate of the cells. This phenomenon was used by Roman to facilitate the isolation of adenine mutants in the elucidation of the adenine biosynthetic pathway: mutations in genes preceding ade_2 in the series would block biosynthesis so that B would not be synthesized. For example, mutation at the ade_1 locus in the diagram would block the step A to B. The result would be a double mutant (ade_1 ade_2) which would still be adenine-requiring but would be non-pigmented and so readily recognizable as a white revertant. By this means it was possible to isolate mutants involved in all the steps in the pathway up to precursor B, each being identified by the appropriate genetic tests.

Mutations resulting in enzyme deficiency are generally recessive, and this is true in the case of ade_2. Thus, in a cross to wild type, the resulting heterozygous (ADE_2/ade_2) diploid would be adenine-independent and non-pigmented. Platings of diploid cells would give rise to white colonies, but following mitotic recombination, homozygous ade_2/ade_2 cells could arise and would be distinguishable as red sectors in the developing white colonies (Fig. 2.4). Red cells can be picked off, subcultured, and sporulated to provide proof that they are diploid. Recessive drug-resistance genes are also useful markers, since homozygous recombinants would be resistant and readily detected.

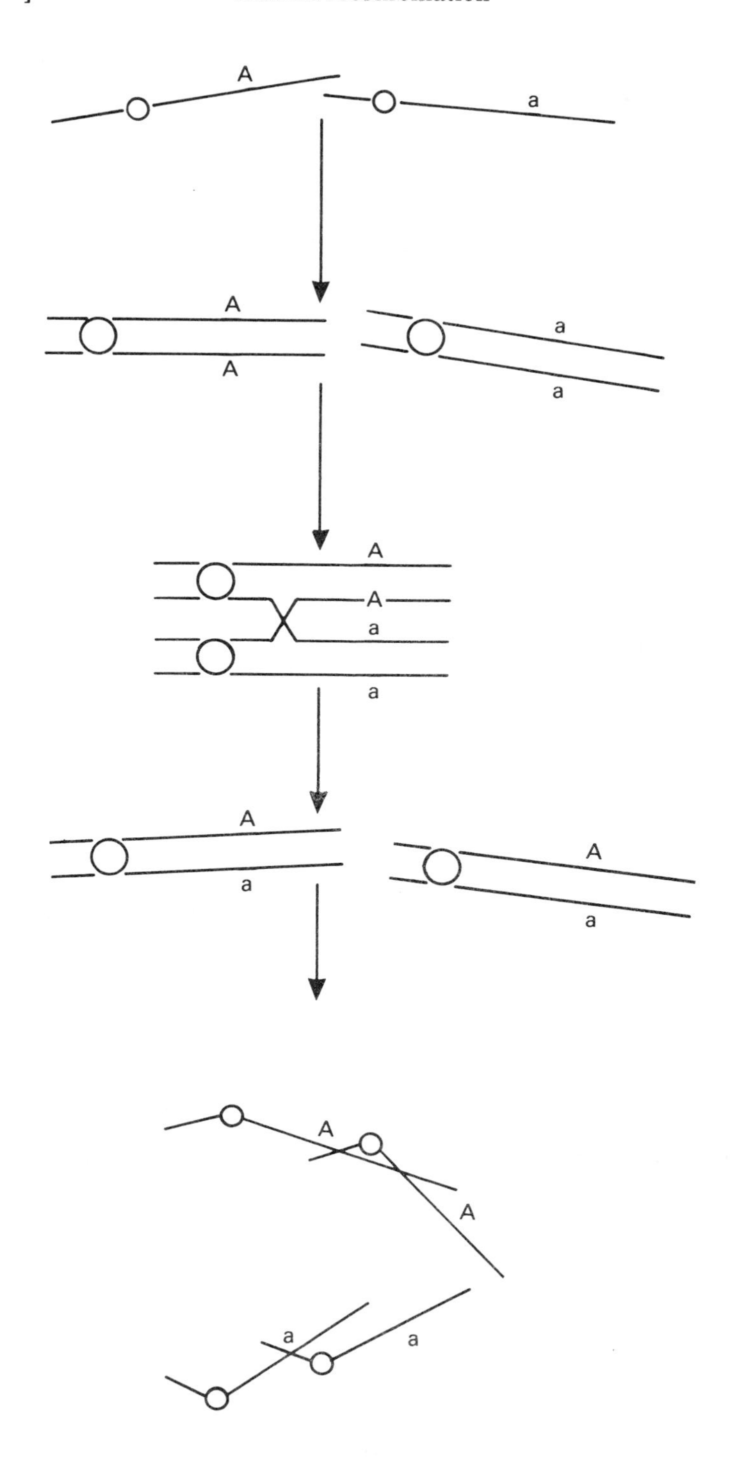

Fig. 2.3 — Diagram of mitotic recombination. See text for details.

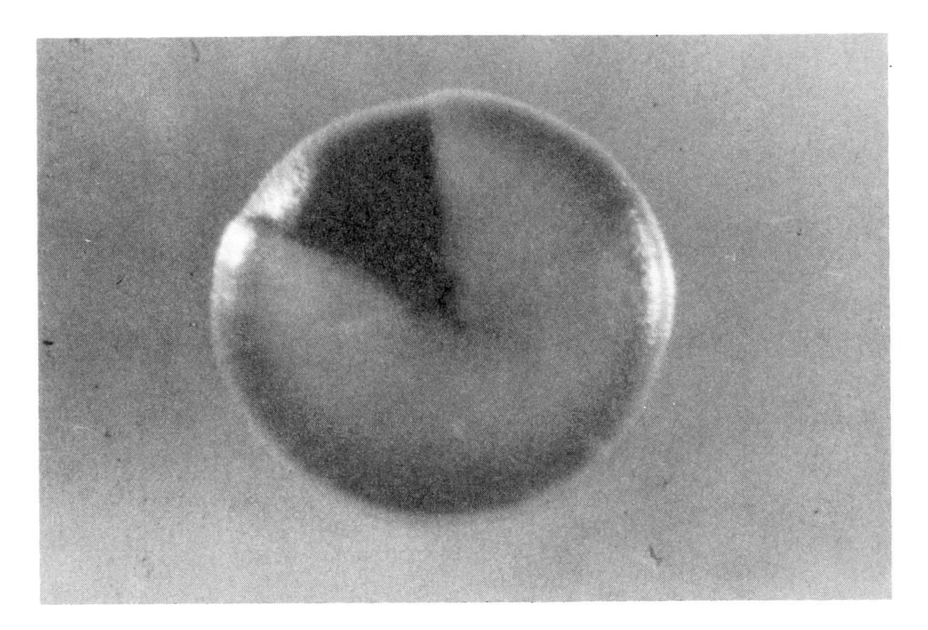

Fig. 2.4 — White, heterozygous diploid colony (*ADE_2/ade_2*) showing homozygous red (*ade_2/ade_2*) sector following mitotic recombination early in colony development.

The phenomenon of mitotic recombination has been exploited in asexual fungi such as Penicilium to obtain recombinants which would not be available except perhaps by more complicated procedures of genetic engineering. Another useful feature of the process is seen in detecting mutagenic potential of drugs, since mutagenic agents can induce the aberration.

2.5 DRUG RESISTANCE MUTATIONS: ISOLATION AND GENETICS

Most drugs and mutagens arrest the growth of yeast cells, but the potency of any drug can vary considerably between strains from little or no effect to strong inhibition. Genetic background is thus of critical importance in drug screening, so it is necessary to use a number of strains to obtain an overall picture of the inhibitory capabilities of any drug. A typcial response to the presence of an inhibitor is shown in Fig. 2.5 in which 22 strains of yeast are used in a series of Petri dish cultures. It can be seen that the range of tolerance to the particular drug under test is wide, with at least a 100-fold difference between strains in tolerance. In further investigation, the most sensitive strains are selected. Resistance mutants are obtained by plating large numbers of cells on drug-containing medium on which they will come up as isolated colonies with a frequency usually of the order of 10^{-5}. In Fig. 2.5, resistant colonies can be seen against a background of inhibited cells

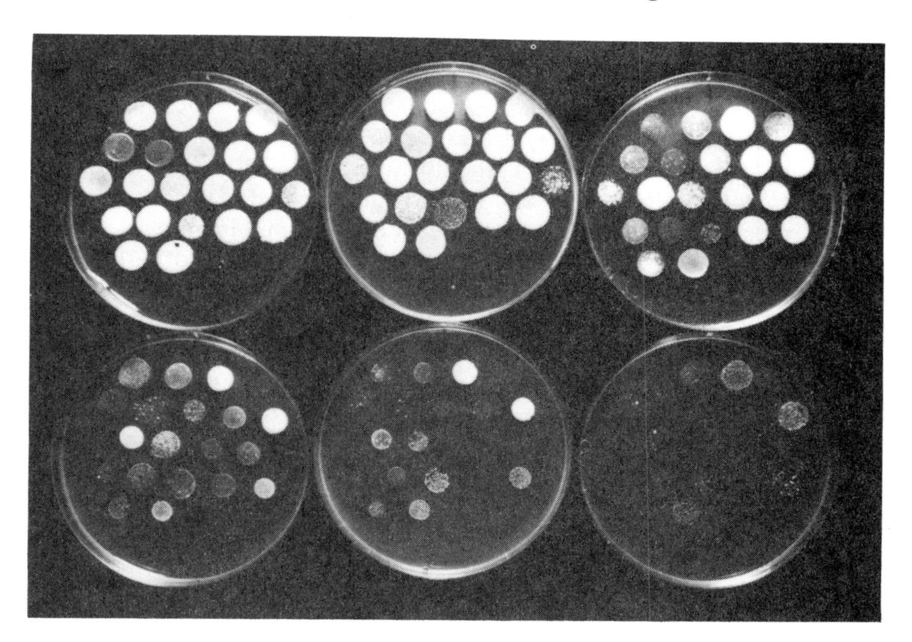

Fig. 2.5 — Drop inocula of 22 haploid yeast strains onto a series of Petri dishes containing chlorimipramine in concentrations 0, 5, 10, 100, 500, and 1000 μg/ml reading from top left to bottom right. Extent of variation in tolerance among strains is typical of most drugs. From Hughes & Wilkie (1972).

from drop inocula. Since these mutants have arisen spontaneously, it may be assumed that a single genetic change is responsible for resistance in each case.

An example of the genetic analysis of drug-resistant mutants is available with respect to cycloheximide which blocks protein synthesis on cytoplasmic ribosomes (80-S type) (Cooper *et al.* 1967). In crosses between resistant mutants and wild type (sensitive), the expected ratio of 2:2, resistant: sensitive was generally seen in meiotic tetrads, indicating single nuclear gene control in each cases. The resistance levels of the individual diploids showed that some of these genes were fully dominant, some semi-dominant, and others recessive, with semi-dominant genes the most frequent. The presence of modifying genes was also detected, i.e. genes which by themselves do not confer resistance but interact with resistance genes to increase or decrease resistance levels (Wilkie & Lee 1965). Modifying genes are indicated when resistance levels of resistant segregants fall into two or more categories. An alternative method to tetrad analysis (which requires micromanipulation) is random spore analysis in which the sensitive parental strain carries the red ade_2 marker. In crossing to a resistant mutant, the diploid obtained would have the genotype $ade_2/ADE_2\ D^s/D^r$ (where D represents the drug resistance gene) and would be phenotypically white, adenine-independent. Diploid cells are then sporulated, treated with glusulase or snail gut juice to digest away the ascus wall, washed, and plated on yeast extract (YED) medium. Non-sporulated diploid cells and ascospores which inherited ADE_2 will give rise to white colonies, while ascopores inheriting ade_2 will give red colonies. These can be picked off and tested for drug resistance. If ade_2 and D are on different chromosomes they will show independent assortment, and the ratio of ade_2 D^S (parental type) to ade_2 D^R (recombinant type) will be 1 : 1. If the two genes are linked there will be a preponderance of parental type, and the recombinants will comprise significantly fewer than 50% of these meiotic progency. The mating type gene will also be segregating so that red, drug-resistant recombinants can be isolated with either mating type. These can be used as standards in crosses to other resistant mutants in establishing allelism or non-allelism of resistance genes. Where resistance is controlled by different genes it may be assumed that the mechanisms of resistance are different. One of these would be an alteration in the binding site so that the mode of action of drugs could be facilitated by using appropriate resistant mutants.

2.6 THE MITOCHONDRIAL SYSTEM

Mitochondria have an intrinsic genetic system based on the possession by the organelle of a single chromosome. In Saccharomyces, as in most eukaryotic organisms, this genome is a circular molecule of double-standard DNA of about 25 μM in contour length (70–80 000 base pairs). A good proportion of this DNA of the yeast mitochondrion in non-coding, comprising extensive sequences of A-T. About 40 or so genes are located in the coding segments, and these specify the tRNAs and the ribosomal RNAs of the mitochondrial protein-synthesizing system and certain polypeptides of the inner membrane assembly of the organelle. The latter comprise 3 subunits of the polymeric cytochrome *c* oxidase (cytochrome aa_3), cytochrome *b*, 2 subunits of the ATPase complex, and var_1, a protein of unknown function. Although the nucleotide sequence has been determined the complete informational content of the molecule is not yet available. The structural genes are transcribed and translated on mitochondrial ribosomes inside the organelle by a regulatory system as yet not fully understood. There is no evidence to show that transcripts of nuclear genes get into mitochondria and are translated there; and, in any case, the two translation systems are incompatible with differences in recognition of certain codons. The mitochondrial protein synthesizing system is further distinct from that of the cytoplasm in that the ribosomes are sensitive to the activity of antibacterial antibiotics whose site of action is the bacterial ribosome (70S type). These drugs include erythromycin, chloramphenicol, and the tetracyclines. Apart from the extensive non-coding segments of the DNA of Saccharomyces mitochondria, the general features of coding capabilities, protein synthesizing system, organization and function are similar in mitochrondria generally from man to fungi.

It is apparent that the mitochondrial genome provides a relatively small number of mitochondrial proteins, and that the bulk of mitochondrial components are coded by nuclear genes the transcripts of which are translated on cytoplasmic ribosomes. It is still not clear how the latter polypeptides get across membranes into the organelle.

2.6.1 Mitochondrial mutants

2.6.1.1 The petite *mutation*

The yeast *Saccharomyces cerevisiae* is a facultative anaerobe and as such can grow without respiratory functional mitochondria. Under anoxia, proteins of the inner membrane assembly are repressed in their synthesis, and ATP is provided by glycolysis which requires a constant supply of glucose or other suitable fermentable energy source. Thus, any aberration, mutation or inhibition of the mitochondrion which precludes respiration would be equivalent to anaerobiosis and would not be lethal or arrest cell growth in fermentable medium. This is certainly true of the so-called *petite* mutation. This is a bizarre event of unknown mechanism in which the mitochrondrial genome is decimated so that only a fragment (frequently not more than 5%) of the mitochondrial DNA (mitDNA) survives for transmission to daughter cells. Occasionally, *petites* arise which have no detectable mitDNA. Daughter cells inheriting such mitochondria are respiratory defective and cannot grow in non-fermentable medium. The massive deletion of information precludes the elaboration of a protein-synthesizing system in the organelle exemplified in the failure of cytochromes aa_3 and b to develop (Fig. 2.6). On the other hand, the synthesis of mitochondrial proteins coded by nuclear genes and assembled in the cytoplasm continues in *petite* cells exemplified in the production of cytochrome *c*. Being deletion mutations, the *petite* condition is irreversible and so is perfectly stable. Another striking feature of the mutation is its high frequency of about 10^{-2} so that 1 in 100 cells in most strains will show the condition. This makes the mutation easy to score since plating of cell samples on YED medium will usually yield one *petite* colony per hundred colonies. The high frequency of spontaneous mutation also makes diagnosis of mitochondrial change quite certain, since nuclear mutations never seem to occur at such a rate but are usually of the order of 10^{-6}.

2.6.1.2 The petite *mutation in mutagenicity and toxicity testing*

Petite colony can be induced by mutagenic agents (Fig. 2.7) and, apparently in many cases, with much greater efficiency than nuclear mutations. Thus, acriflavin and ethidium bromide can achieve up to 100% induction of *petite* depending on the strain without having much effect on nuclear change using lethal mutations as a criterion. When known chemical carcinogens were tested (Table 2.1), there was induction of *petite* and the selective nature of the mitochondrial mutation was established on 4 counts: (1) there was little or no increase in reversion to prototrophy in strains carrying biochemical markers such as histidine auxotrophy at the concentrations of carcinogen used to induce *petite*; (2) no significant effect on cell viability at these concentrations; (3) mitochondrial but not cytoplasmic protein synthesis was depressed or arrested as seen from absorption specta in which cytochrome *c*

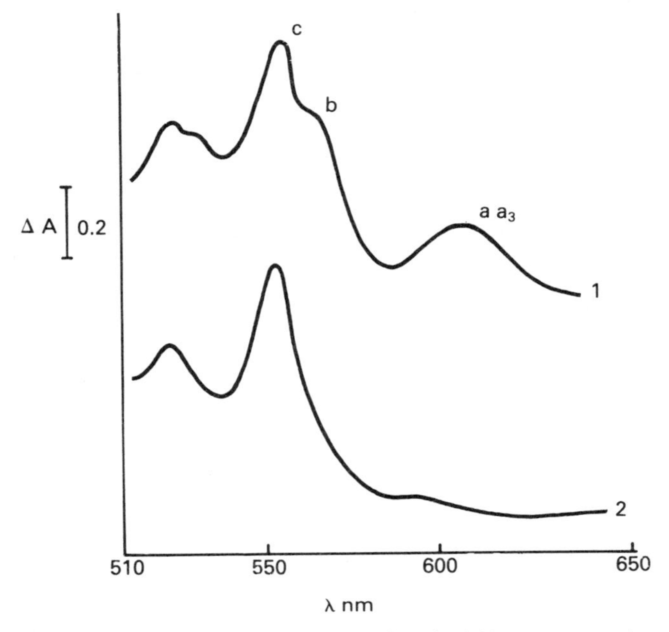

Fig. 2.6 — Absorption spectra of cells of strain A32 grown to stationary phase in YED medium. 1, wild type cells; 2, cells of *petite* mutant peaks of cytochromes aa_3, *b*, and *c* occur at 603, 562, and 550 nm respectively with mutants; peaks α β peaks of *b* and *c* at 530 and 520 nm respectively. (Uvispek 1800 SP recording spectrophotometer. Cell density 5×10^8 per ml).

but not aa_3 and *b* developed (Fig. 2.6); (4) inhibition of growth of cells in non-fermentable YEG medium but not in the glucose-containing YED medium (Egilsson *et al.* 1979).

Selective reactivity of chemical carcinogens with mitDNA has also been demonstrated in animals. For example, mitDNA extracted from liver cells of mice treated with radioactive hepatocarinogens (aromatic hydrocarbons) showed up to 400 times the radioactivity of nuclear DNA from the same cells (Allen & Coombes 1980); mitDNA from the liver of rats treated with dimethylnitrosamine was alkylated much more extensively than the nuclear DNA from the same tissue (Wilkinson *et al.* 1975); hamster cells cultured in carcinogens of various chemical affinities showed significantly greater reactivity of these compounds with the mitDNA of these cells than nuclear DNA with differences up to 100-fold (Backur & Weinstein 1980).

The reason for the greater mitochondrial sensitivity may lie in the different organization of the mitochondrial chromosome which, unlike nuclear chromosomes, is a closed circle of 'naked' DNA with little if any packaging protein. This feature may render mitDNA more easily accessible to DNA-reacting compounds; but, on the other hand, thre are a good many drugs that are not mutagenic, do not induce the *petite* mutation, but have mitochondria as primary targets (Wilkie 1971). This suggests that perhaps

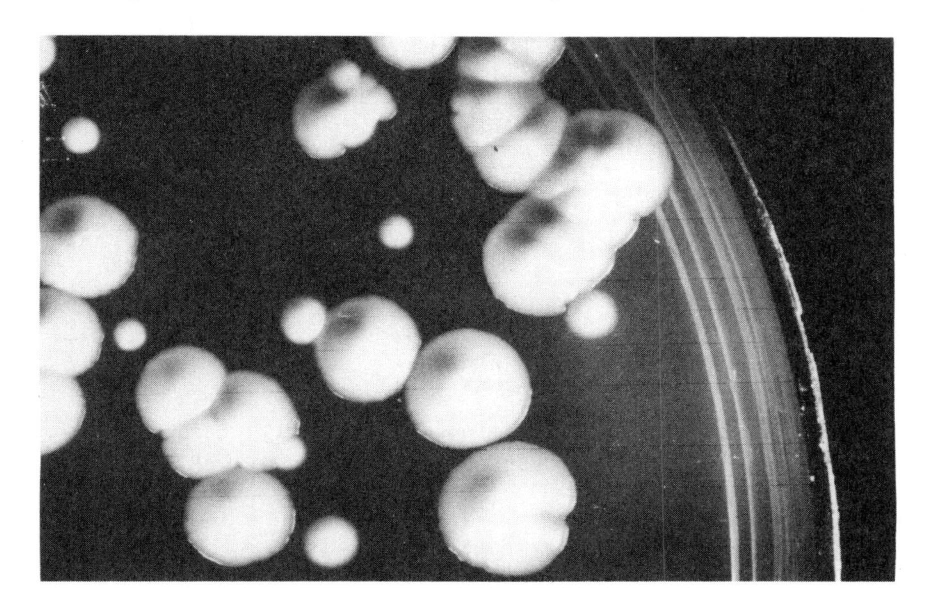

Fig. 2.7 — The mitochondrial mutant *petite* (small colonies) with normal colonies growing on YED agar medium. The high frequency of *petite* in this culture follows induction with thioacetamide.

the organelle preferentially accumulates foreign compounds generally, and consequently it is at this site that the greatest activity is seen. Whatever the mechanism of nitochondrial reactivity, induction of the *petite* mutation would seem to be a useful criterion for screening drugs and chemicals for potential carcinogenicity.

The first indication of primary antimitochondrial activity of a drug and the most easily scored is inhibition of growth of yeast cells in non-fermentable medium while growth continues in glucose medium in the presence of the drug. The next step would be to separate the antimitochondrial agents into the two categories of inhibitors of biogenesis and inhibitors of function. DNA-reacting compounds would come into the former class as would the ribosome-reacting antibacterial antibiotics, and would be detected as inducers of *petite* and/or inhibitors of mitochondrial protein synthesis. Inhibitors of function would usually be membrane-reacting drugs such as the tricyclic antidepressants (Wilkie 1971) and dinitrophenol inhibiting respectively the respiratory chain and oxidative phosphorylation, effects which are reversible. It is possible, once having detected primary antimitochondrial activity, to assess toxicity in terms of the production of toxic-free radicals. Since mitochondrial blockage by genetic damage or otherwise is not lethal to cells in glucose medium, any killing effect must be due to extra-mitochondrial cellular damage. However, cell death is not often due to the direct action of the drug molecule itself but rather to toxic free radicals into which drugs are converted following their metabolism by the cell's mixed function

Table 2.1 — Effects of chemical carcinogens on growth and frequency of the mitochondrial *petite* mutation in strains of *Saccharomyces cerevisiae*.

MIC_G* ($\bar{x}$)	$MIC_{D/G}$ ($\bar{x}$)**		Highest*** $MIC_{D/G}$	*Petite* induction
A. *Carcinogens*				
4-nitroquinoline-N-oxide	0.40 ug/ml	2.17	6.67	+ + +
Ethionine	1.30 ug/ml	6.35	13.33	—
Adriamycin	15.0 ug/ml	13.4	55.0	+ +
Diaminobenzidine	8.0 ug/ml	5.0	15.0	+
Dichlorobenzidine	28.0 ug/ml	2.49	4.5	—
Cadmium chloride	63.6 ug/ml	1.73	10.0	+
2-Naphthylamine	65.9 ug/ml	4.25	10.0	+
9,10-Dimethyl-anthracene	210.0 ug/ml	3.8	3.0	—
Ca chromate	56.2 ug/ml	4.2	10.0	+
Ethylene thiourea	401.0 ug/ml	24.34	>24.9	+ +
Cobalt chloride	640.0 ug/ml	1.32	2.0	+
Benzidine	1.3 mg/ml	2.62	4.0	+ + +
Thioacetamide	0.55 mg/ml	18.6	36.0	+ +
Thiourea	0.64 mg/ml	7.8	> 7.81	+ +
Nickel chloride	3.34 mg/ml	1.36	2.33	+
Streptozotocin	7.5 mg/ml	>10.0	>10.0	+
B. *Non-carcinogens*				
1-Naphylamine	200 ug/ml	1.25	1.25	—
Anthracene	893 ug/ml	1.34	2.0	—
Tetramethyl-bernzidine	150 ug/ml	1.33	1.33	+ + +

* Minimum inhibitory concentration to arrest growth in YEG medium (non-fermentable) $\bar{x}$, average of MICs from an average of 14 strains tested.
** Ratio of MICs on glucose (YED) medium to MICs on glycerol medium, averaged.
*** Ratio for the strain sensitive on YEG also showing the widest discrepancy between YED and YEG.

oxidase system, generating at the same time, activated oxygen species mainly superoxide and the hydroxyl radical (O_2^- and OH^-). In turn, cells have a system which can regulate at least to some extent, the levels of intracellular free radicals, and it is based on the glutathione peroxidase pathway (Fig. 2.8). In common with other organisms, yeast cells produce glutathione constitutively which scavenges free radicals. When drugs are in excess, an exogenous supply of glutathione can overcome residual toxicity.

The procedure is illustrated in our study of Adriamycin (Nudd & Wilkie, 1983). This anthracycline antibiotic is of clinical importance as a broad spectrum anticancer drug, but is restricted in its use by a dose-dependent cardiotoxicity associated with prolonged treatment by the drug. The ability of Adriamycin to accept electrons from microsomal and nuclear flavoproteins resulting in the formation of a semiquinine free radical and ultimately in the formation of superoxide and hydroxyl free radicals, has been demonstrated in animal cells and in isolated cardiac submitochondrial particles (Backur, 1979). The drug has been shown to intercalate DNA, and is

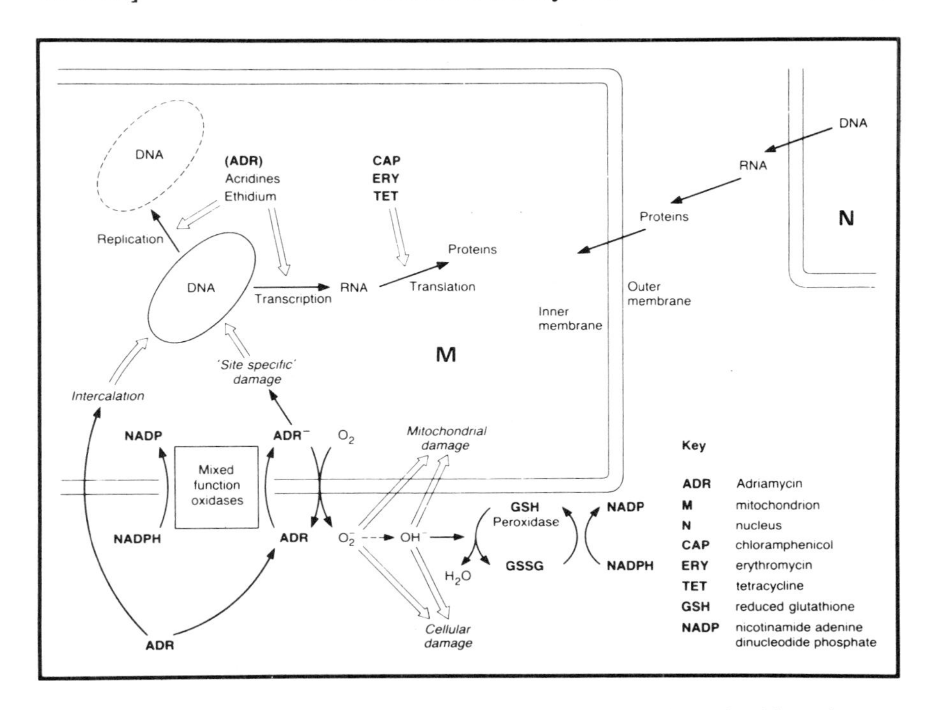

Fig. 2.8 — Anti-mitochondrial activity and oxidative metabolism of Adriamycin. From Nudd & Wilkie (1983).

included in our list of mutagens/carcinogens which block mitochondrial biogenesis and induce the *petite* mutation (Table 2.1). The semiquinone free radical may also intercalate DNA causing 'site specific' damage, and/or may be recycled continually generating a large number of oxygen-centred free radicals (Fig. 2.8). Other studies in animal systems have shown that Adriamycin depletes the levels of reduced glutathione (GSH) and that cardiac cells are more susceptible to sequestration of the anthracycline than other tissues. There are a number of reports that other sulphydryl donors can reduce the toxicity of Adriamycin, and conversely, pretreatment with diethyl maleate which depletes cellular GSH, potentiated the lethality of Adriamycin. Our results with yeast cells parallel these findings in that the killing effect of the drug was largely reversed by an exogenous supply of glutathione but was increased in the presence of diethylmaleate (Fig. 2.9). Furthermore, toxicity (lethality) was significantly enhanced in yeast cultures when they were exposed to phenobarbitone at the same time as Adriamycin (Fig. 2.9). This is explainable on the basis of the barbiturate's known activity as an inducer of mixed function oxidases which would enhance the rate of oxidative metabolism of Adriamycin. The general conclusion that the toxicity of this anticancer drug was due in large measure to the production of oxy-radicals seemed well founded. On the other hand, the DNA reactivity of Adriamycon (scored as *petite* induction) was not significantly affected by

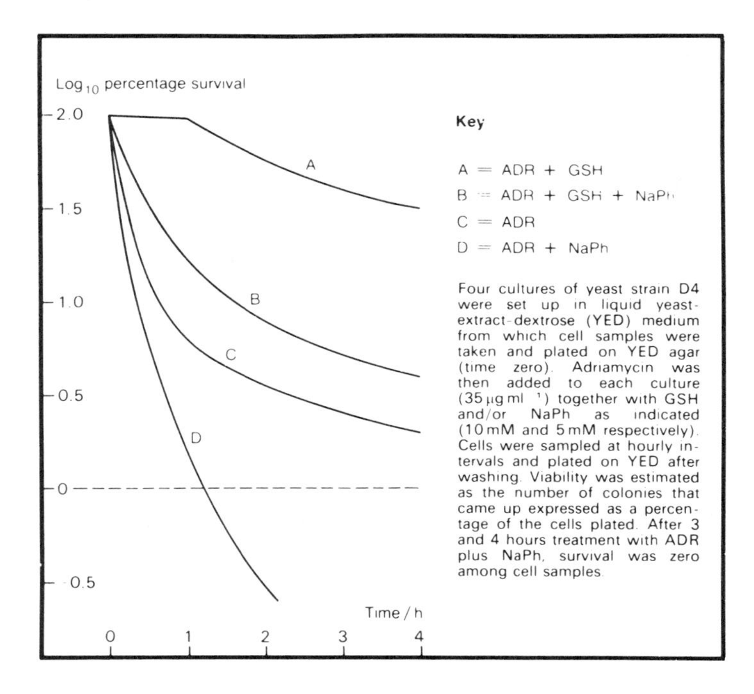

Fig. 2.9 — Toxicity of Adriamycin: reversal by glutathione and potentiation by phenobarbitone (NaPh). From Nudd & Wilkie (1983).

the presence of glutathione, diethylmaleate, or phenobarbitone, indicating that mutagenicity was due mainly to the activity of the intact drug molecule and that free radical metabolites were less important in this respect in this case. If the anticancer efficacy of Adriamycin is in large measure due to DNA reactivity it would seem that toxic effects may be alleviated by administering antioxidants without prejudice to clinical usefulness. Vitamin E is a powerful antioxidant and may be more readily available than glutathione for administration to patients. In our yeast system, the vitamin was only slightly less effective than GSH in reversing Adriamycin toxicity.

It is clear that the yeast system can contribute to the problem of drug toxicity testing by providing a rapid method for assessing toxicity in terms of the production of free radical metabolites. In those cases where drugs have primary antimitochondrial activity, results can indicate whether oxidative metabolites are more efficient than the parent compound, or, indeed, whether their formation is essential before any biological activity is seen. That a connection exists between anti-mitochondrial activity and clinical potential seems reasonable in the case of anticancer drugs, assuming DNA reactivity is a main factor in drug action. In the case of membrane-reacting, non-mutagenic agents, it is not clear whether primary antimitochondrial activity has any connection with the pharmacological effects. However, after classifying a number of derivatives and structural analogues of the lipophylic tricyclic antidepressant imipramine on the basis of their primary targeting of mitochondrial respiration a good correlation was seen between this effect and clinical potency (Wilkie 1971).

2.6.1.3 The mitochondrial system and tetracycline antibiotics

Another attribute of the yeast mitochondrial system is seen in the analysis of antibacterial antibiotics which inhibit protein synthesis in bacteria at the site of translation on the 70-S type ribosome. As already mentioned, these antibiotics specifically inhibit mitochondrial protein sysnthesis (Fig. 2.6), establishing an affinity between the two systems at least with respect to this ribosomal characteristic. Tetracycline comes into this category, and there are a number of reports on the compartive potencies of various derivatives of this antibiotic in bacteria. We undertook a study of the comparative antimitochondrial activity of some of these to test how close or otherwise was the relationship between the two systems. If a positive correlation was found, the mitochondrial activity could be used to classify tetracyclines and possibly the other antibiotics as well which come into the same category. In any case, information on antimitochondrial activity would also be of use in the assessment and study of toxic side effects of these drugs, assuming that the organelle is the primary site of action. In view of the high degree of specificity of the antibiotics for the mitochondrial target, this is not an unreasonable assumption.

The tetracycline formulations that were compared were: tetracycline (TC), oxytetracyline (OTC), chlortetracycline (CTC), desmethylchlortetracycline (DMCTC), 6-methyleneoxytetracycline (MOTC), and tetracycline-L-methylene-lysine (TCML). Primary antimitochondrial activity of these compounds was established using standard procedures. The first of these was inhibition of growth of strains in non-fermentable (YEG) medium at concentrations which allowed growth to proceed when glucose (YED) medium was used. The second criterion was failure of cytochromes aa_3 and b to develop in cells growing in YED medium (Fig. 2.6). As can be seen from the results listed in Table 2.2, minimum inhibitory concentrations (MICs) required to arrest growth in YEG varied with the strain. It can also be seen that there is a good cross-correlation between compounds within each strain. Thus, strains relatively sensitive to TC were sensitive to the derivatives, while strains resistant to high concentrations of TC were relatively resistant to the others. No strain used in these tests was inhibited in glucose medium even at the highest concentration of 5 mg/ml which was near saturation. The cross-relationship was underlined by the analysis of more than 100 spontaneous resistant mutants to TC. In every case, resistance to TC was accompanied by increased tolerance to the TC derivatives. These results provide good evidence that these compounds have the same site of action in mitochondria, so that comparing their relative potencies in blocking the reactive site seems a valid exercise.

From the results given in Table 2.2, TC, OTC, and CTC have similar inhibitary capabilities with CTC perhaps marginally more efficient than the other two. DMCTC is clearly the most potent of the derivatives, being as

Table 2.2 — Minimum concentrations of tetracycline (TC) antibiotics inhibiting growth of yeast strains utilizing non-fermentable substrate.

	Minimum inhibitory concentrations (mg/ml)*					
Strain	TC	OTC	CTC	MOTC	DMCTC	TCML
A10	0.025	0.05	0.025	0.05	0.025	0.1
A12	0.05	0.05	0.05	0.05	0.025	0.1
A31	0.05	0.05	0.05	0.1	0.05	0.25
A7	0.1	0.1	0.05	0.1	0.05	0.25
A8	0.1	0.1	0.05	0.25	0.05	0.75
A13	0.1	0.1	0.05	0.25	0.05	0.25
A30	0.1	0.1	0.05	0.25	0.05	0.25
A22	0.25	0.25	0.25	0.5	0.1	0.75
A26	0.25	0.25	0.25	0.5	0.25	0.75
A32	0.25	0.25	0.25	0.25	0.1	0.75
A34	0.25	0.25	0.25	0.5	0.1	0.75
A5	0.05	0.5	0.5	0.75	0.25	1
A11	5	5	5	>5	2.5	>5
A15	5	5	5	>5	2.5	>5

* Range of antibiotic concentrations used: 0.01, 0.025, 0.05, 0.01, 0.25, 0.5, 0.75, 1, 2.5, 5 mg/ml.

TC, tetracycline; OTC, oxy-TC; CTC, chlor-TC; MOTC, 6-methylene-oxy-TC; DMCTC, demethyl-chlor-TC; TCML, TC-6-methylene-lysine.

active as CTC in half of the strains and more active in the others. MOTC is comparable to TC in the more sensitive strains but less active than TC in strains that have a relatively higher level of tolerance to TC. TCML is the least potent of all and only in two strains is this derivative comparable to MOTC. These findings with respect to anitmitochondrial activity were compared with results from bacterial studies. Comparative potencies of drugs in bacteria are usually presented as percentage of strains inhibited at a given concentration of antibiotic. This is useful in giving a general picture of comparative antibacterial activity, but in the present context it is more useful to have data on individual bacterial strains to compare with our results for mitochondrial inhibition. A comprehensive study of TC derivatives in this respect is that of Last *et al.* (Last *et al.* 1969), using *Escherichia coli* as the test organism. They compared the potencies of TC, OTC, and CTC in inhibiting (1) growth, (2) peptide synthesis *in vitro*, (3) binding of aminoacyl-tRNA to ribosomes *in vitro* (*in vitro* in this connection means cell-free system). Firstly they observed a correlation, as others have, between relative activity *in vitro* and against bacterial growth, and secondly that the three antibiotics were similar on all counts in potency. The only exception was that CTC was 0–20% more effective than TC and OTC in the inhibition of peptide synthesis.

A number of laboratory appraisals have been made of the inhibition of *in vitro* growth of bacteria by DMCTC compared with other derivatives. Many organisms have been screened, and, in general, results show that a significantly higher proportion of strains are more sensitive to DMCTC than to TC, OTC, and CTC, the last three being in the main, rather similar in potency (Steibigel *et al.* 1968). In similar studies (English *et al.* 1968), MOTC was marginally better than or of equal potency to DMCTC, which in turn was more potent than TC against Gram-positive bacteria. With Gram-negative bacteria, the order of potency tended to be DMCTC, TC, MOTC in descending order. As far as we are aware, published data are not available of the comparative efficiency of TCML against *in vitro* bacteria, and the manufacturers (Erba) were not able to provide references. To this end, we recorded the inhibitory effects on growth of a culture of *E. coli* strain JG151, in the presence of TCML at various concentrations. From optical density readings, percent inhibition was calculated and compared with that obtained for TC (Fig. 2.10). It is clear that TCML is significantly less potent than TC

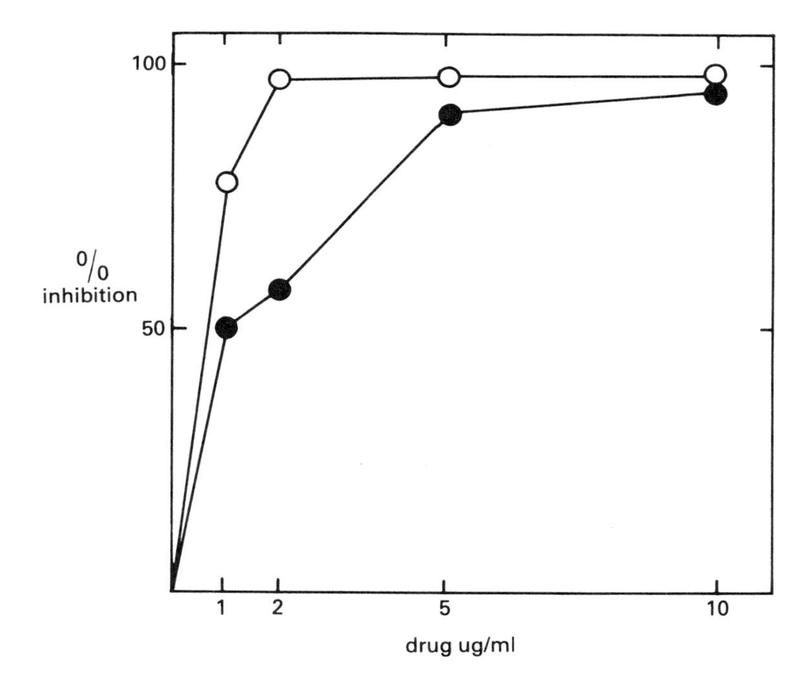

Fig. 2.10 — Inhibition of growth of *E. coli* by tetracyclines in oxoid nutrient broth. ○—○, tetracycline; ●—●, tetracycline-6-methylene lysine.

in arresting growth of this strain of *E. coli*.

In summarizing, a correlation seems to exist between antibacterial and antimitochondrial activity, particularly if the bacteria used for comparison are of the coliform type, and the general conclusion that TC derivatives that are potent inhibitors of bacteria are likely to be potent in antimitochondrial activity, is not unwarranted. This generalization depends on the validity of extrapolation of results with yeast to mitochondria of other organisms. In view of the striking similarity in structure and function of all mitochondria, and the finding that bacterial antibiotics of the above type similarly affect yeast and mammalian mitochondria, this is not unrealistic.

It is apparent from these studies that the yeast mitochondrial system can be exploited in classifying drugs which primarily attack mitochondria in terms of mode of action and aspects of toxicity. These drugs would include not only antibacterial antibiotics but also certain lipophyllic agents, many anticancer drugs, and most carcinogens.

2.6.2 Mitochondrial genetics

2.6.2.1 The petite *mutant*

Once established in a cell, the *petite* mutation is transmitted during growth and cell division thereafter to all vegatative progeny. When a *petite* mutant is crossed to wild-type, normal strains, the resulting diploid clones comprise diploid cells which, when sampled and plated, give rise to a mixture of *petite* colonies and normal colonies on YED agar medium (Fig. 2.7). The ratio of *petite* : normal colonies varies; depending on the strains used, but can be as high as 9:1 in some cases. These results indicate that when *petite* and normal cells fuse to form zygotes, a mixture of *petite* and normal mitochondrial types is established in the zygote cytoplasm. During growth and development of zygote clones, *petite* and normal mitochondria segregate out among the vegetative progency so that, early in clone development, individual diploid cells inherit one or other mitochondrial type, and a heterogeneous mixture of mitochondria is not maintained for long. The implication is that, although there may be many mitochondrial genomes in a cell, only very few are available for transmission to daughter cells.

When normal colony segregants are sporulated, all meiotic progeny inherit normal mitochondria, and tetrads show 4 normal : 0 *petite*. *Petite* diploid colonies cannot sporulate, and so cannot contribute further to the genetic analysis. The segregation of the two mitochondrial types during cell division (not to be confused with mitotic recombination which occurs at a much lower frequency and has consequences for linked genes) and no segregation in meiosis clearly shows a departure from the rules governing the inheritance of nuclear genes.

2.6.2.2 Mitochondrial antibiotic resistance

Specific inhibition of mitochondria by the antibiotics precludes the growth of cells inoculated onto YEG medium, but when cells are plated in sufficient numbers, spontaneous resistant mutant cells will give rise to colonies with a frequency of about 10^{-5}. Since there are mitochondrial genes which specify components of the mitochondrial ribosome, it was not unreasonable to expect that the mechanism of resistance in some of these mutants would result from mutation in one or other of these genes, causing altered component and so affect drug binding. Using the criteria of vegetative segregation and non-Mendelian ratios in tetrads as in *petite* inheritance, mitochondrial resistance mutations were in fact identified. Thus, in a cross between a mitochondrial erythromycin resistant (ERY^R) mutant and an erythromycin sensitive strain (ERY^S), diploid clones comprised cells which had segregated for ERY^R and ERY^S. When ERY^R diploid segregants were isolated and sporulated, all meiotic progeny were ERY^R, and tetrads showed 4 ERY^R to 0 ERY^S. Conversely, the meiotic haploids from ERY^S segregants were all ERY^S, and tetrads showed 4 ERY^S : 0 ERY^R. The conclusions to be drawn from these findings were the same as for *petite* inheritance, namely that zygotes contained a mixture of mitochondrial types which segregated out among diploid daughter cells. Once segregation had occurred only one mitochondrial type was transmitted thereafter, whether to sexual (meiotic) or asexual (mitotic) progeny.

A third criterion was used in locating mitochondrial drug resistance mutations. Since it was known that *petite* mutants had suffered extensive loss of their genetic information, it could be assumed that *petite* mutants of ERY^R strains would have lost the resistance gene in most cases. This was proved to be the case when these *petite* ERY^R strains were crossed to the normal ERY^S strains to yield diploid progeny which segregated for *petite* and normal as expected but which showed no inheritance of ERY^R among the normal colonies (it is not possible to check for ERY tolerance in *petite* cells since these do not grow in YEG medium at the outset). Occasionally, a particular *petite* of an ERY^R mutant does transmit the resistance gene, in which case it may be concluded that the particular segment of mitDNA remaining in the *petite* genome contained the resistance gene. There was thus the obvious possibility of building up a bank of *petites* each containing a specified gene, the only limitation being the number of marker genes available.

Using these criteria for locating mitochondrial genes, other antibiotic resistance mutants were isolated and identified, including resistance to chloramphenicol (CAP^R), tetracycline (TC^R), spiramycin (SPY^R), and oligomycin (OLI^R) pincipally. When crosses were set up between strains with different resistance genes, not only did parental types segregate out but recombinant types were also identified among the diploid progeny. Thus, in a cross between strains carrying respectively ERY^R and CAP^R (i.e. ERY^R CAP^S × ERY^S CAP^R), classes of diploid segregants were obtain, namely, ERY^R CAP^S and ERY^S CAP^R the 2 parental types and ERY^R CAP^R and ERY^S CAP^S the 2 recombinant types. In crosses involving 3 resistance genes, the 8 possible classes emerged in diploid clones and in 4-point crosses, all 16 assortments of genes were obtained. To obtain recombinant types, crossing over between the different mitDNAs must have taken place, and the relative frequencies of recombinants allowed rough linkage maps to be constructed. This exercise was complicated by the high frequency of recombination between marker genes irrespective of their order along the mitochondrial chromosome, giving rise to the theory that multiple exchanges occurred between mitDNAs along the lines of a viral cross. How mitDNA molecules from different mitochondria come together to exchange segments is not clear, but a model based on mitochondrial breakdown and realease of the molecules in zygotes has been proposed (Smith *et al* 1972).

2.6.2.3 Episome-like inheritance of TC-resistance: a special case

Although mutations in mitochondrial genes were identified using the above procedures, most resistance mutations were nuclear in the control of antibiotic resistance i.e., no segregation during clonal development of diploids from zygotes in crosses to sensitive strains and 2:2 segregations of R:S in tetrads. A special case arose in which a TC-resistance mutation showed both a cytoplasmic and nuclear mode of transmission. (Hughes-Wilkie 1972), depending on whether cells were in mitotic or meiotic division. The first criterion of mitochondrial location of the TC^R gene was met when in the cross to TC^S, diploids were seen to segregate for TC^R and TC^S. When the TC^S segregant was sporulated and tetrads analysed, the second criterion was at least partially realized in that ratios of 4 TC^S : 0 TC^R were obtained. However, unexpectedly, diploid cells which had inherited TC^R gave tetrad ratios of 2 TC^R : 2 TC^S, indicating nuclear inheritance of TC^R. A backcross was then made between a TC^R meiotic product and the TC^S parental strain, and once again the pattern of inheritance was seen, namely, segregation among diploid progeny, failure of TC-resistance to be transmitted either mitotically or in meiosis in TC^S segregants, and 2:2 segregation of TC^R:TC^S in tetrads from TC^R diploids.

The conclusion drawn from these results was that perhaps this TC^R gene was not located in mitochondria in vegetative cells but was to be found as a genetic element in the cell cytoplasm somewhat analogous to an R-factor in bacteria. Zygotes from the cross $TC^R \times TC^S$ would have the factor, and diploid daughter cells inheriting it would show TC resistance while those failing to do so (and this would be a random process) would be TC sensitive as would be their meiotic progeny. The model further postulated that in TC^R diploids going through a meiotic division, the genetic element carrying TC^R became integrated in a nuclear chromosome to be replicated and transmitted in meiotic division, 2 products of the tetrad inheriting the integrated chromosome and 2 inheriting non-integrated homologues. Finally, on zygote formation the factor would be once again released into the cytoplasm to be inherited at random then stabilized in diploid cells. This was thought to be an early example of an episome-type transmission of antibiotic resistance in yeast. Alternatively, it is still possible that the TC^R gene had a mitochondrial location in which it was transmitted during vegatative growth and division, but was translocated to the nucleus during or at the onset of meiosis. This hypothesis may have some support in recent reports of the presence of sequences of mitDNA in nuclear chromosomes (Fox, 1983).

2.6.2.4 Physical evidence of mitochondrial recombination

The appearance of recombinant types and their segregation pattern in diploids provide circumstantial evidence that crossing-over takes place between mitochondrial chromosomes. Direct evidence of this process was obtained from a study of the inheritance of cleavage sites of endonucleasus in mitDNA. Firstly, it was established that the fragments generated from the mitDNAs of two strains A and B by the restriction endonucleases *Hpa*II and *Hae*III differed in the location of some cleavage sites. These enzymes split the DNA wherever the respective sequences C–C–G–G and G–G–C–C appear, and essentially break down the mitochondrial genome into its constituent genetic units. Electrophoretic separation an agarose/polyacrylamide gels of the DNA fragments (roughly 100) gave a fragment pattern characteristic of each strain but with approximately a 10% difference: certain fragments appeared in the digests of strain A which were not present in strain B and *vice versa*, indicating this degree of difference in the location of cleavage sites along the polynucleotide chains. A and B were then crossed and individual diploids isolated from which mitDNAs were extracted and subjected to digestion with *Hpa*II and *Hae*III. Separation on gels (Fig. 2.11) showed recombination of restriction sites in that each fragment pattern was characterised by fragments originating from each parent, but there were also new fragments to be seen. (Fonty *et al.* 1978). The first finding provides by itself unequivocal evidence of physical recombination of the parental genomes. The finding of new fragments in these diploid segregants gives an important clue to the mechanism of the recombinational process since they probably resulted from unequal exchange. The most likely location of such events is in the extensive A-T regions (spacers) where internal sequence repetition could provide opportunities for illegitimate pairing.

2.6.2.5 Mitochondrial-nuclear interactions

The mitochondrial genome has a coding capacity limited to about 40 genes. These specify components of the organellar protein-synthesizing machinery (tRNAs and rRNAs) and perhaps a dozen or so polypeptides. The latter are assembled on the organellar ribosomes and contribute to the differentiated structure of the inner membrane. There is no evidence that any mitochondrial gene products find their way out of the organelle to be incorporated into other cellular systems. The remainder (probably more than 90%) of mitochondrial components are coded by nuclear genes and assembled on cytoplasmic ribosomes. When mitochondrial protein synthesis is blocked or disrupted as in *petite* mutants or cells growing in the presence of the antibacterial antibiotics, the nuclear-coded organellar components continue to be made, as exemplified in the production of cytochrome *c*. It would appear then that interaction between the two genetic systems is unidirectional from nucleus to mitochondria, and that the organelle plays no part in regulating the activity of the nuclear genes. Recent evidence indicates that this statement may require modification.

As well as causing respiratory deficiency the mitochondrial mutation to the *petite* condition has a pleiotropic effect on characterizatics of the plasma membrane (Evans & Wilkie 1983). These include alteration in cellular agglutinability with the lectin concanavalin A, failure to take up certain metabolites, alteration in tolerance to drugs, loss of flocculence in flocculating strains (Fig. 2.12), changes in cell morphology, and cell surface electro-

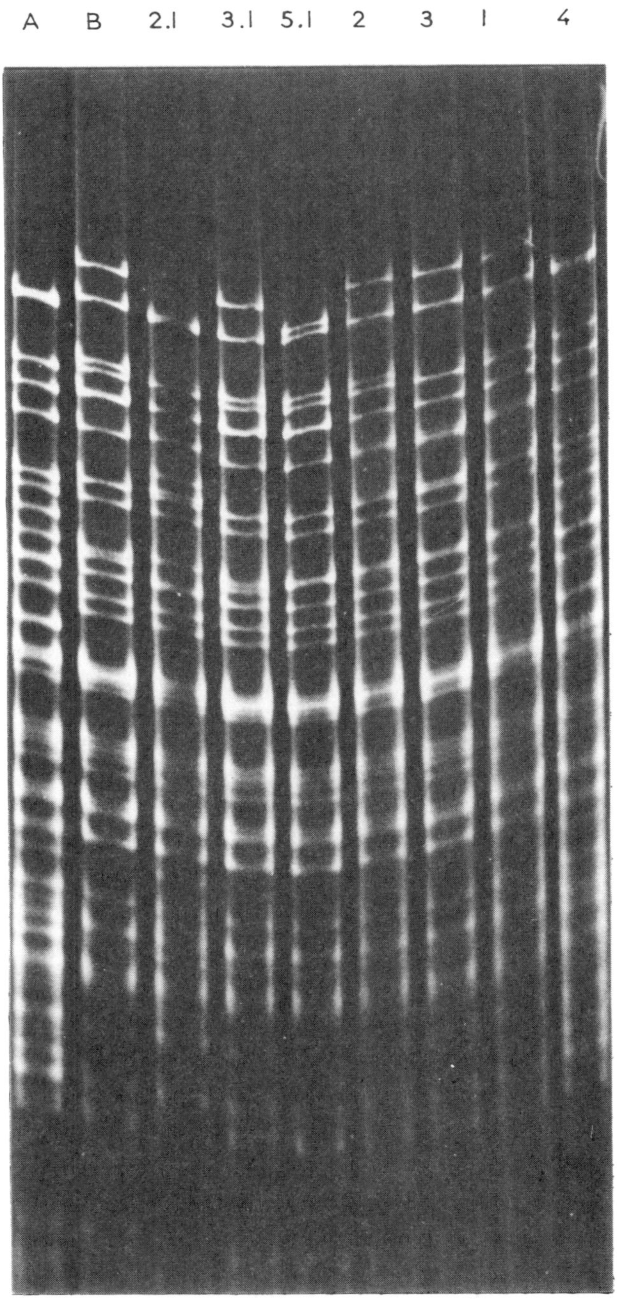

A B 2.1 3.1 5.1 2 3 1 4

Fig. 2.11 — Electrophoretic patterns on 0.5% agarose/2% polyacrylamide gel of *Hae* digests of mitDNAs from 2 haploid parental strains of yeast A and B and purified diploid cultures obtained by plating cell samples from zygotes produced by matings of A and B. From Fonty *et al.* 1978.

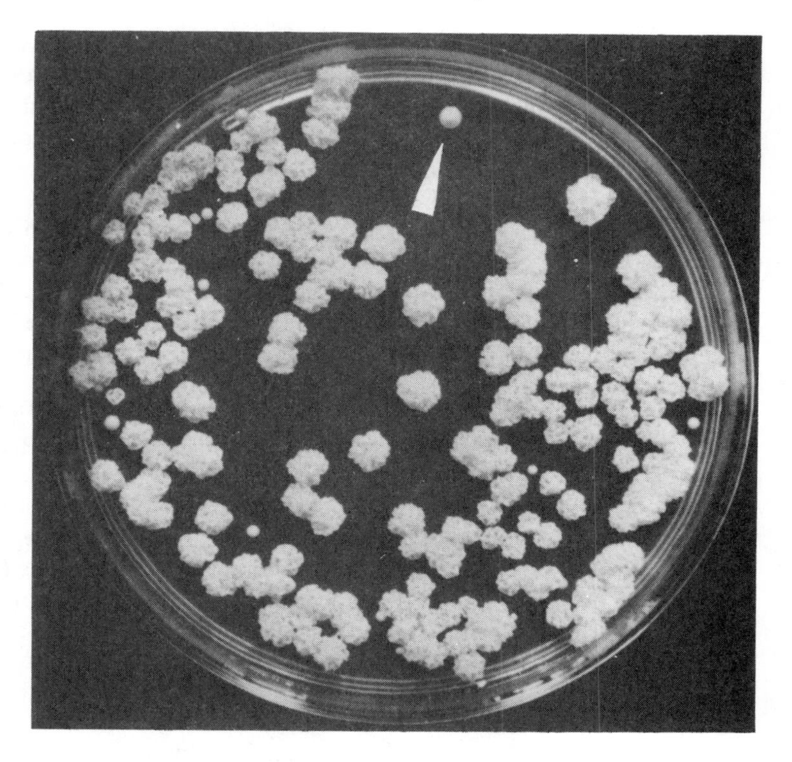

Fig. 2.12 — *Petite* (smooth) colony in a flocculent strain of yeast which has rough colony morphology due to cell clumping (surface aberration). From Evans & Wilkie (1983).

kinetic properties. *Petite* mutants of all our strains show these changes, but the extent of the changes seen depends on the strain. Thus in strain D22, the *petite* mutant has a reduced ability to take up the sugar galactose, and growth is slow when this sugar is the sole carbon and energy source. On the other hand, *petite* mutants of strain D6 are totally incapable of taking up this sugar and do not grow. If cells of the latter *petite* mutant are permeabilized with DMSO, galactose will be taken up and the appropriate glucosidase will be induced (Mahler & Wilkie 1978), demonstrating that the aberration is indeed cell surface related. The problem then centres on how the functional state of mitochondria can affect components of the plasma membrane, particularly certain minor proteins (Wilkie & Nudd 1981).

Genetic analysis of the strain differences with respect to the performance of their respective *petite* mutants provide important clues. For example, in the case of galactose utilization (gal$^+$) *versus* non-utilization (gal$^-$), a cross was set up between strains D22 and D6. Meiotic tetrads were cultured from each of which *petite* mutants were isolated and tested for galactose utilization. The tetrads of *petites* tended to segregate mainly 2 gal$^+$: 2 gal$^-$, indicating that a nuclear gene or genes controlled the strain differences. In other words, the extent to which the *petite* condition modified cell surface characteristics depended on which alleles of certain genes were present in the nucleus, leading to the general conclusion that mitochondria had a role in modulating the activity of certain nuclear genes. Presumably, these genes were specifying the cell surface components, and there is the possibility that mitochondria function in modifying the products of these genes prior to their incorporation into the plasma membrane. The flux of Ca^{2+} in cells is largely under mitochondrial control, and since this divalent cation is frequently involved in the activation of proteins, organellar control of this process could be a possible mechanism. Whatever the mechanism, the evidence gives a clear indication that mitochondria have a wider role in eukaryotic cells than providing ATP in energy metabolism, and that there is a two-way interaction between mitochondrion and nucleus.

A further example of this interaction is seen in the case of mitochondrial antigens of unknown function in primary biliary cirrhosis (PBC). Patients showing the condition produce antibiotics in their serum against these antigens (Klitakin & Kantor 1972). Antigens which react with these PBC antibodies have been detected in yeast mitochondria, but they also have an extra-mitochondrial location, probably the plasma membrane (Ghadiminejad 1986). Strains produce the anitigens when their mitochondrial protein synthesis is blocked either by growth of cells in erythromycin or by the *petite* condition, indicating synthesis takes place in the cytoplasm from transcripts of nuclear genes. Genetic evidence from crosses between strains with respectively high and low levels of antigen production also indicates control by nuclear genes. However, repression of mitochondrial enzymes (including cytochrome *c*) by high glucose concentrations, anoxia, or continuous exposure of cells to dinitrophenol during growth (DNP reacts with the mitochondrial inner membrane with a high degree of specificity), prevents synthesis of the antigens. It may be speculated from these findings that the PBC antigens come into the category of mitochondrial proteins coded by nuclear genes which are repressible. When the genes are active, the antigens require a particular configuration or state of the mitochondrial inner membrane for their final assembly. When fully formed, a proportion of these units migrate to the plasma membrane where, if there are cell surface aberrations, improper exposure may stimulate an immune reaction. Whatever the mechanism, these findings once again demonstrate the principle of mitochondrial involvement in the expression of nuclear genes.

A third and final example of this phenomenon is the depression of the synthesis of catalase in *petite* mutants compared with the amounts in parental strains (Evans & Wilkie (unpublished)). Once again nuclear genetic background is an important factor since the degree of repression of the enzyme by the *petite* condition is strain-dependent and varies from a small effect to failure to detect any catalase activity. Since the enzyme is linked to oxidative metabolism in the production of activated oxygen species, a mitochondrial link is not unexpected. However, the mechanism of the mitochondrial effect regulating the nuclear coded enzyme, as in the cases described above, remains an open question.

2.7 DISCUSSION

The nuclear and mitochondrial genetic systems of Saccharomyces are probably the most extensively studied of all eukaryotic microorganisms. This is not surprising in view of the overall ease of handling of these organisms as demonstrated herein. These systems can be exploited in problems relating to the classification of drugs, to their modes of action, and to aspects of their toxicity. The selective mitochondrial mutagenicity of chemical carcinogens may have implications for the process of oncogenesis itself, particularly since mitochondrial aberrations can cause changes at the cell surface, considered by many to be a main step if not a prerequisite for the onset of neoplasia. It may be necessary, then, to reconsider in the light of the evidence presented here, the old hypothesis of Warburg that mitochondrial mutagenesis may yet be another route to the cancerous condition. The yeast cell is also an ideal vehicle for the study of mitochondrial–nuclear interactions, particularly the fundamental problem of the regulatory role of the organelle in nuclear gene activity. In view of the fact that mitochondria of all organisms are similar in all fundamental respects, results obtained with yeast cells can be extrapolated to humans with some measure of confidence.

REFERENCES

Allen, J. A. & Coombes, M. M. (1980) *Nature* **287** 244.

Backer, J. M. & Weinstein, I. B. (1980) *Science*, **209** 297.

Backur, N. R. (1979) *Proc. Natl. Acad. Sci. Wash.* **76** 297–304.

Cooper, D., Banthorpe, D. V., & Wilkie, D. (1967) *J. Mol. Biol.* **26** 347–350.

Egilsson, V., Evans, I. H., & Wilkie, D. (1979) *Molec. Gen. Genet.* **174** 39–46.

English, A. R., McBride, J. F., & Riggio, R. (1968) In: *Antimicrobial agents and chemotherapy* pp 462–472 (M. Finland & G. M. Sanger, eds) Amer. Soc. Microbiol., New York.

Evans, I. H. & Wilkie, D. (Unpublished results).

Evans, I. H. & Wilkie, D. (1983) *Trends in Biochem. Sci.* **7** 147–151.

Fonty, G., Goursot, R., Wilkie, D., & Bernaddi, G. (1978) *J. Mol. Biol.* **119** 213–235.

Fox, T. D. (1983) *Nature* **301** 371–372.

Horn, P. and Wilkie, D. (1966) *J. Bacteriol.* **91** 1388–1389.

Hughes, A. R. & Wilkie, D. (1972) *Heredity* **28** 117–127.

Klitskin, G. & Kantor, F. S. (1972) *Ann. Intern. Med.* **77** 535–541.

Last, J. A., Izaki, K., & Snell, J. F. (1969) *Biochim. Biophys. Acta* **174**, 1959–1960.

Mahler, H. R. & Wilkie, D. (1978) *Plasmid* **1** 125–133.

Nudd, R. & Wilkie, D. (1983) *Chem. Brit.* **19** 911–915.

Smith, D. G., Wilkie, D., & Srivastava, K. C. (1972) *Microbios* **6** 231–238.

Steibigel, N. H., Read, C. W., & Finland, M. (1968) *Amer. J. Med. Sci.* **255** 179–184.

Stern, C. (1936) *Genetics* **21** 625–630.
Wilkie, D. (1971) In: *Cellular toxicity of anesthetics* pp 160–167 (B. R. Fink ed.) Williams & Wilkins, Chicago.
Wilkie, D. & Nudd, R. C. (1981) In: *Vth International Symposium on Yeasts* (G. G. Stewart & I. Russell, eds), Pergamon Press, Toronto.

APPENDIX

Culture media

YED, 1% Difco yeast extract, 2% glucose
YEG, 1% yeast extract, 4% glycerol
Sporulating medium, SPM: sodium acetate, 0.3%, raffinose, 0.04%
Minimal medium: 10% mineral salts solution (see below), 0.1% vitamin solution (see below), 0.1% $CaCl_2$ solution (10%), glucose 2%.

Mineral salts solution:

H_2O	1 litre
$(NH_4)_2SO_4$	10 gm
KH_2PO_4	8.75 gm
K_2HPO_4	1.25 gm
$MgSO_4.7H_2O$	5 gm
NaCl	1 gm
H_3BO_3	0.1 ml of 0.1% solution
$CuSO_4.5H_2O$	0.1 ml of 0.1% solution
KI	0.1 ml of 0.1% solution
$FeCl_3.6H_2O$	0.1 ml of 0.5% solution
$ZnSO_4.7H_2O$	0.1 ml of 0.7% solution

Vitamin solution:

H_2O	100 ml
Biotin	0.2 mgm
Thiamine	40.0 mgm
Pyridoxine	40.0 mgm
Inositol	200.0 mgm
Ca pantothenate	40.0 mgm

3

Mechanisms of gene induction and repression in *Saccharomyces cerevisiae*

Dr. Peter W. Piper
Department of Biochemistry, University College London

Abbreviations

UAS:	upstream activator sequence
ORF:	open reading frame
ARS:	autonomously replicating sequence
Kd:	kilodaltons
bp:	basepair
nt:	nucleotides

3.1 INTRODUCTION

Since the advent of recombinant DNA technology we have begun to understand how *S. cerevisiae* genes are controlled at the molecular level. Most of this control has been found to operate at the level of transcription. This chapter considers the mechanisms and controls known to act during the transcription of yeast nuclear protein coding genes by RNA polymerase II and during the ensuing maturation of functional mRNAs. The synthesis of rRNAs, tRNAs, and mitochondrial RNAs is not discussed.

The intensive research over the past eight years into the structure and function of *S. cerevisiae* genes could not have been performed without the development of ways of introducing DNAs constructed *in vitro* back into yeast cells for functional analysis (Section 3.3). There are as yet no yeast transcription systems comparable to the mammalian systems (Manley 1983) that faithfully reproduce transcription initiation by RNA polymerase II *in vitro*. Clearly for detailed biochemical studies an *in vitro* transcription system will have to be developed, but in the meantime the ease of genetic manipulations and DNA transformations in yeast make its sequences readily amenable to functional analysis *in vivo*.

One of the major driving forces behind the recent expansion of research into yeast gene expression has been the desire to use *S. cerevisiae* as a host for the expression of heterologous proteins. Several such proteins have been synthesized in *S. cerevisiae* (reviewed by Kingsman *et al,* 1985), and some such as tissue plasminogen activator (tPA) and the surface antigen of hepatitis B virus (HBsAg) are currently undergoing clinical trials. It is to laboratories interested in the biotechnological applications of yeast that we owe much of our knowledge of the necessary requirements for high-level intracellular synthesis, regulated synthesis, or secretion of proteins in this organism. *S. cerevisiae* is capable of secreting active antibody (Wood *et al.* 1985) and producing HBsAg particles indistinguishable from those in the serum of patients who are hepatitis B virus carriers (Valenzuela *et al.* 1982). Yeast has tremendous potential as an expression system, subject to two limitations. Firstly the removal of intron sequences in yeast transcripts through RNA splicing is more sequence-stringent than in higher eukaryotes (Section 3.2.3). Thus introns need usually to be removed from a mammalian coding sequence before it can be expressed in yeast using a yeast promoter. Secondly yeast does not glycosylate proteins in the same way as mammalian cells, and is not therefore suitable at the present time for the synthesis of mammalian proteins whose biological activity requires complex glycosylation (generally proteins which, unlike tPA, have a long half-life in the circulation).

3.2 THE SYNTHESIS AND MATURATION OF TRANSCRIPTS MADE BY RNA POLYMERASE II IN *S. CEREVISIAE*

3.2.1 The organization of protein coding genes and mRNA

The events leading to the synthesis of mRNA and protein from a typical intronless yeast gene are summarized in Fig. 3.1. Although most of the genes of *S. cerevisiae* lack introns (alternatively called intervening sequences or I.V.S.s) some essential ones do have them. The removal of introns through splicing is a possible target of control mechanisms which act only upon intron-containing genes and is considered in section 3.2.3.

A gene is by definition a functional entity, and as such comprises both those sequences that encode protein (protein coding region or open reading frame (ORF)) and the flanking DNA essential for the correct expression of this region. It is usual to talk about those flanking sequences to the 5′ side of the ORF on the coding strand as upstream DNA and those sequences to the 3′ side of the ORF as downstream DNA (Fig. 3.1). Upstream sequences are especially important in protein gene expression as they contain the transcriptional promoter (section 3.4), the region of the DNA where RNA polymerase II binds and separation of DNA strands occurs with formation of the open-promoter complex, prior to the initiation of RNA synthesis. Also some of the promoter sequences, especially the TATA box (section 3.4.2), position the single or multiple sites of transcription initiation. Various regulatory sequences are also located within the upstream regions

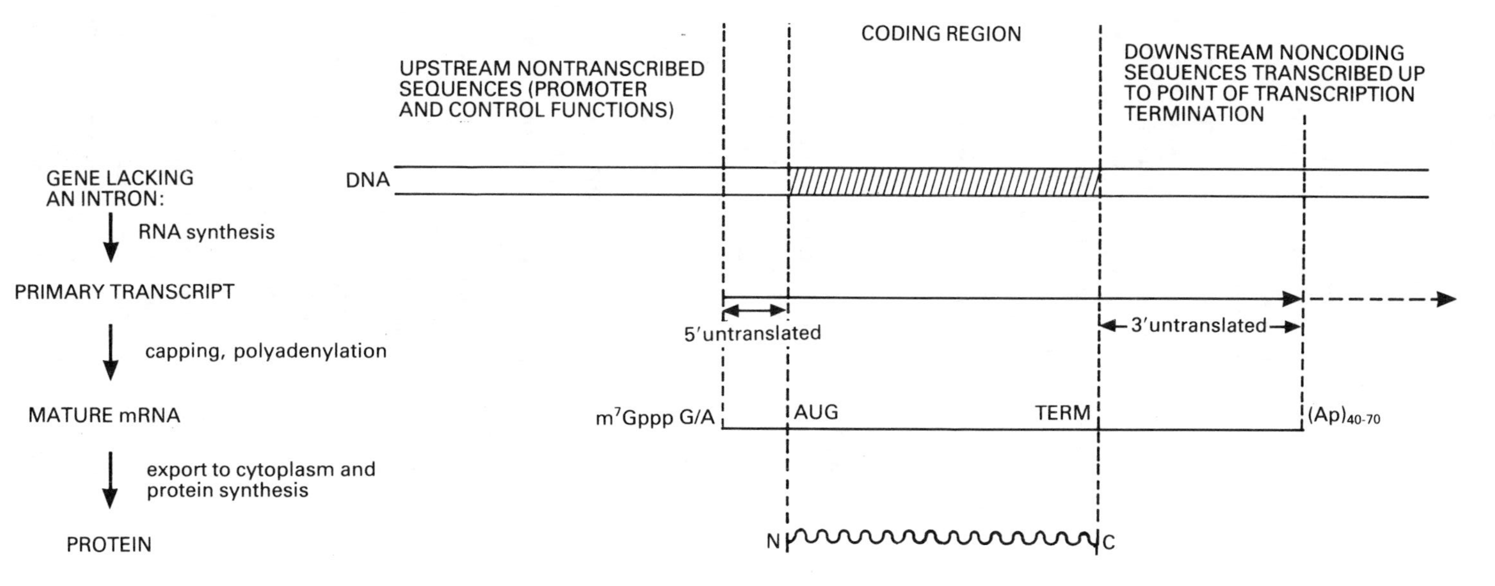

Fig. 3.1 — The events leading to expression of an intronless protein gene. DNA sequences upstream of the actual protein coding region (hatched) contain a promoter for RNA polymerase II as well as sequence elements that operate as a part of control systems governing the level of expression of the gene. All these events, apart from protein synthesis, occur in the nucleus.

of eukaryotic genes and, by mechanisms that are only beginning to be understood at the molecular level (sections 3.4.2, 3.5, and 3.6), act as part of control systems to influence promoter efficiency.

3.2.2 Synthesis of 5′ and 3′ termini of mRNA

RNA synthesis initiates downstream of promoter sequences a variable distance, usually less than a hundred nucleotides, upstream of the AUG methionine initiation codon of the ORF (Figs 3.1 and 3.2). It continues to a transcription termination site located somewhere within the 3′ noncoding region. The 5′ terminal base of yeast RNA polymerase II transcripts is a purine which is found modified as either a $m^7G(5')ppp(5')G$ or a $m^7G(5')ppp(5')A$ cap structure in mature mRNAs (Sripati *et al.* 1975). In higher eukaryotes there is evidence that 5′ capping may be an integral part of transcription initiation (Darnell 1982), and while this probably also applies in yeast it has yet to be proven experimentally. Both the cap structure and the 5′ untranslated region of an mRNA appear to be essential for efficient ribosome binding and initiation of protein synthesis (Darnell 1982). The 5′ untranslated or 'leader' regions of most *S. cerevisiae* mRNAs do not share an obvious homology to the 3′ end of 18S rRNA that could comprise a ribosome-binding site analogous to the Shine–Dalgarno sequence of prokaryotes (Shine & Dalgarno 1975). However, in some individual mRNAs a homology with rRNA has been noted (Zalkin & Yanofsky 1982). Translation usually commences at the first AUG methionine codon of the mRNA and may also require certain structural features in its immediate vicinity since the base three nucleotides before this AUG is generally an adenine (Dobson *et al.* 1982). The mRNA leader may also be important in the selective translation of certain mRNAs after heat shock (McGarry & Lindquist 1985).

The sequences downstream of the ORF of yeast genes are generally A+T rich and contain signals both for the termination of transcription upon the DNA and the polyadenylation of the transcript. Polyadenylation is the post-transcriptional addition of a short stretch of polyadenylic acid (poly(A)) to the 3′ end of the mRNA during maturation. Unlike in higher eukaryotes (Darnell 1982), all yeast mRNAs appear to have 3′ poly(A). However, the yeast mRNA tails comprise only some 40–70 A residues, and are therefore shorter than the poly(A) tails of higher eukaryotes (100–200 A residues). Yeast has mRNAs of both relatively short half-life (of 15–20 min) and of long half-life (of 60–70 min) (Santiago *et al.* 1986). The length of the poly(A) tail is the same for both populations of mRNA (Piper, unpublished results). It has not yet been conclusively demonstrated whether yeast poly(A) tails are added to the 3′ end of nascent transcripts or to termini generated by endonucleolytic cleavage (Figs 3.1 and 3.2). In higher eukaryotes transcription termination usually occurs downstream of the poly(A) addition site, the latter arising through a concerted cleavage and polyadenylation about 15–20 nt downstream of a AAUAAA recognition sequence on the transcript (Darnell 1982). The poly(A) of several yeast mRNAs is found just downstream of a sequence: UAAUAA(A/G), and this may be the

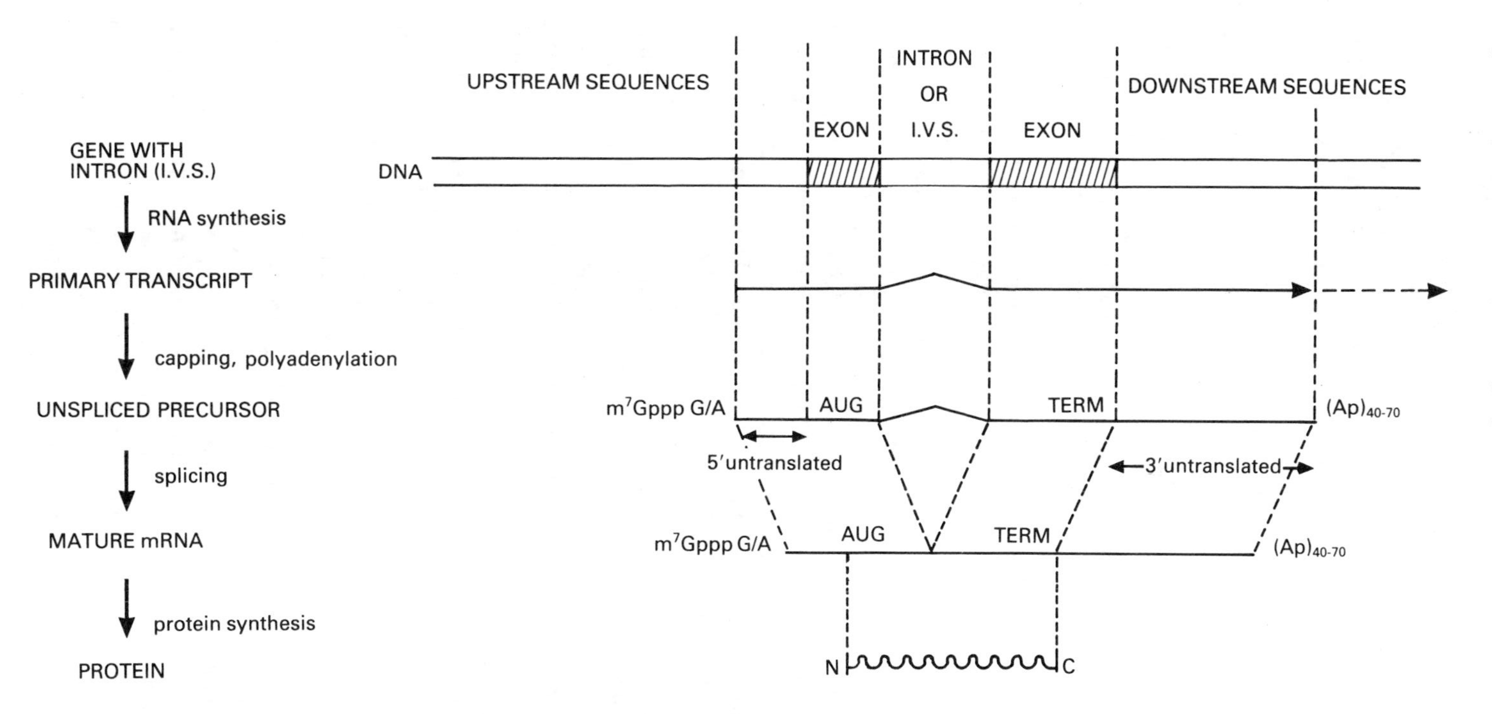

Fig. 3.2 — Expression of a gene in which an intron, or intervening sequence (IVS), disrupts the protein coding region (hatched). The intron sequence is removed during RNA processing prior to export of the mature mRNA from the nucleus to the cytoplasm.

functional analogue of the AAUAAA of higher eukaryotes (Bennetzen & Hall 1982). Deletion of a specific downstream sequence in the *CYC*1 gene abolishes correct transcription termination (Zaret & Sherman 1984), and a similar sequence: TAG . . . TAGT . . . (A/T)rich. . . TT has been identified in several *S. cerevisiae* genes. Analysis of revertants of this 3′ *CYC*1 deletion mutant has provided evidence that, in contrast to the usual situation in higher eukaryotes, transcription termination and polyadenylation may be coupled processes in yeast (Zaret & Sherman 1984). The signals for RNA polymerase II termination and polyadenylation on yeast genes may therefore be identical. A part of these signals may be the sequence: TTTTTATA which is found 50–90 bp upstream of the poly(A) site in many yeast genes and is required for transcript termination upon the *Drosophila ADE*8 gene when this is introduced into *S. cerevisiae* (Henikoff *et al.*, 1983).

3.2.3 The removal of introns through splicing

Although introns are not as widespread in the genes of *S. cerevisiae* as in genes of most other eukaryotes, they are nevertheless present in many vital protein coding regions such as those for certain ribosomal proteins, actin and *MAT*α1. The enzymatic removal of intron sequences from the RNA transcripts of such genes is termed splicing (Fig. 3.2). Introns of nuclear mRNA precursors extend from a GU at the 5′ splice site to an: AG at the 3′ splice site. Comparison of intron sequences in *S. cerevisiae* nuclear protein coding genes revealed an internal conserved 7 nt element: TACTAAC a short distance from the 3′ splice site. This element is essential for proper processing (Langford & Gallwitz 1983; Pikielny *et al.* 1983, Langford *et al.* 1984). A similar, though much less stringent, sequence occurs in other eukaryotic introns, e.g. in *Neurospora:* CT(A/G)AC (Kinnaird & Fincham 1983) and in mammals: CTGAC (Keller & Noon 1984). The TACTAAC box shows striking sequence complementarity to the extreme 5′ end of *S. cerevisiae* introns, suggesting that these sequences base–pair during splicing. Lariat forms of the intron sequence are detectable amongst the transcripts of a number of different nuclear protein coding genes (carried on high copy number plasmids (section 3.3.2) to ensure over-expression). In these lariat intermediates of splicing an RNA branch is formed through cleavage at the 5′ splice site and reattachment of the G at the 5′ end of the intron to the third A (underlined) in the TACTA<u>A</u>C box, via a 2′–5′ phosphodiester bond. This A is thereby participating in both 2′–5′ and 3′–5′ phosphodiester linkages (Domdey *et al.* 1984). The intermediate formed by this reaction contains two RNAs, the 5′ exon as a linear molecule and the intron-3′ exon as a lariat. Subsequent reactions, giving rise to the ligated exons and an excised lariat intron, complete splicing.

The *S. cerevisiae* splicing mechanism is strikingly similar to that in metazoan cells. However, both the branch sites and 5′ splice sites of yeast introns are much more sequence specific than in more complex eukaryotes. The stringent requirement for a TACTAAC box close to the 3′ end of the intron for splicing in *S. cerevisiae* is the probable reason that *S. cerevisiae* is incapable of correctly removing introns from most transcripts of higher eukaryotic genes (Beggs *et al.* 1980, Watts *et al.* 1983, Langford *et al.* 1983), while introns of yeast genes are spliced in extracts from mammalian cells (Sharp 1985). Higher eukaryote genes must therefore have their introns removed at the level of DNA if they are to be expressed in *S. cerevisiae.* In *Schizosaccharomyces pombe* introns are found in genes more frequently than in *S. cerevisiae,* and the conserved sequences in the vicinity of intron–exon boundaries are more similar to those found in higher eukaryotes (Kaufer *et al.* 1985).

Conditional lethal (temperature-sensitive) mutations in the *RNA*2–*RNA*11 genes of *S. cerevisiae* cause, at the nonpermissive temperatures of 30–36°, loss of translatable mRNAs for intron-containing nuclear genes. This is accompanied by increases in the steady-state levels of intron-containing pre-mRNAs (Rosbach *et al.* 1981, Teem *et al.* 1983). Clearly, splicing can be a potential target for the selective control of intron-containing genes. It is doubtful that it plays such a complex role in yeast as in mammalian cells, where control over alternative splicing patterns appears to be important in hormonal or cell type control over gene expression (Sharp 1985). Nevertheless splicing appears to be important in the process whereby *S. cerevisiae* coordinates accumulation of ribosomal proteins with ribosome assembly. Excess ribosomal protein gene transcripts synthesized in cells on high copy number vectors are not spliced, and excess ribosomal protein is degraded rapidly, whereas most other transcripts and proteins accumulate roughly in proportion to gene dosage (Warner *et al.* 1985).

3.3 SYSTEMS FOR IDENTIFYING TRANSCRIPTIONAL CONTROL ELEMENTS

3.3.1 Transformation by exogenous DNA

The elements responsible for transcriptional control upon yeast nuclear protein genes (sections 3.4.2, 3.5, and 3.6) are: (i) regulatory DNA sequences (*cis*-acting elements) closely linked to the gene and usually residing upstream of the coding region; and (ii) proteins (*trans*-acting elements) encoded by loci not generally linked to the genes they control. The *trans*-acting elements are identified by genetic analysis of extragenic control-modifying mutations, and are proteins that frequently directly influence transcription of the genes under their control by binding to the *cis*-acting DNA regions adjacent to these genes. Regulatory DNA sequences are most readily investigated by constructing promoter fusions and gene fusions (Fig. 3.3) and then assaying their function *in vivo.* In several studies hybrid yeast promoters have been constructed consisting of the controlling upstream activator sequence (UAS) of one gene and the TATA box of another. Upon reintroduction into yeast these hybrids direct transcription initiation at sites specified by the TATA box region and are regulated in a way determined by the UAS. Experiments of this kind have shown that there is some flexibility in the spacing of UASs and TATA regions within promoters (section 3.5).

The ability to measure the activity of promoter and gene fusions in

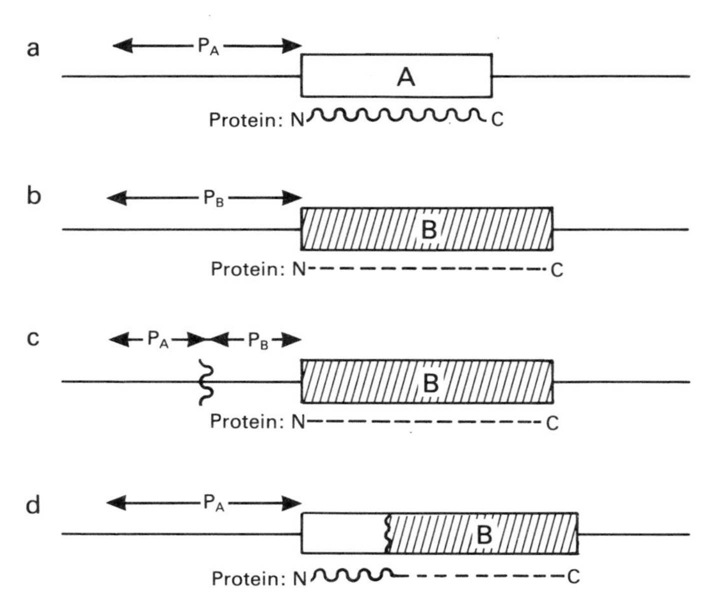

Fig. 3.3 — A diagram to illustrate the difference between promoter and gene fusions. (a): gene A with its promoter sequences (PA) and a protein coding region (open box) oriented by N (amino-terminus) and C (carboxy-terminus). (b): gene B with promoter sequences (PB) and protein coding region (hatched box). (c): a hybrid gene (promoter fusion) produced by joining portions of the promoter sequences from A and B. (d): a hybrid gene (gene fusion) produced by fusion of the coding regions of A and B.

S. cerevisiae cells was made possible by the development of techniques (reviewed by Struhl 1983) by which exogenous DNA can be introduced into yeast cells. Two principal methods of transforming yeast with DNA are currently in widespread use. The first involves enzymatic removal of the cell wall to yield sphaeroplasts, which are then treated with polyethylene glycol and calcium ions in the presence of the transforming DNA, prior to being allowed to regenerate cell walls (Beggs 1978, Hinnen *et al.* 1978). The frequency of transformation by this procedure is high, many plasmid copies being taken up by each cell. The second method of transformation is much less efficient, does not involve cell wall removal, and uses lithium ions to stimulate uptake of exogenous DNA (Ito *et al.* 1983). A genetic marker must be used to detect DNA uptake. Usually an auxotrophic strain is employed, and the wild-type gene which will complement its genetic deficiency is incorporated into the exogenous DNA. On selective media transformants can then be picked from the background of nontransformed auxotrophs. The *LEU*2 (Beggs 1978), *URA*3 (Hinnen *et al.* 1978), and *TRP*1 (Struhl *et al.* 1979) genes are often used as selectable markers together with strains having a deletion at the chromosomal copy of the same gene (this maximizes strain stability as there is no chromosomal site for integration by homologous recombination). Dominant selectable genes (listed in Kingsman *et al.* 1985) which confer drug resistance can also be employed to select transformants, and have the advantage that they can be employed for prototrophic strains.

3.3.2 Relicative vectors

Among the plasmids designed to assay the function of *in vitro* constructed DNAs in *S. cerevisiae* the most versatile are those able to replicate and be selected in both *E. coli* and *S. cerevisiae*. Three kinds of these 'shuttle vectors' are shown in Fig. 3.4 All have convenient restriction sites (not

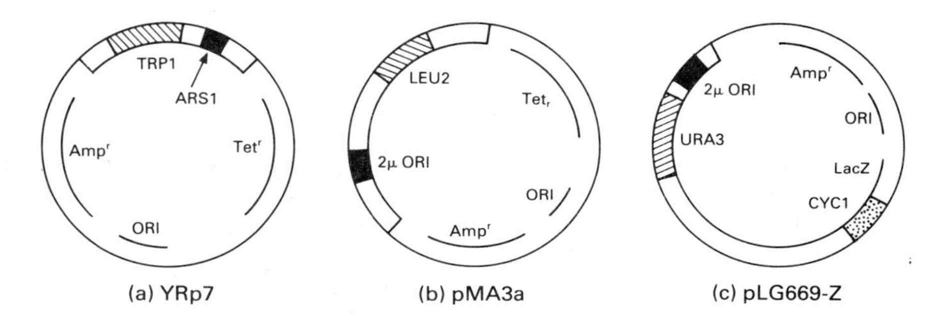

Fig. 3.4 — *E. coli*/*S. cerevisiae* shuttle vectors (not drawn to scale). Portions of sequence derived from *E. coli* plasmids are represented by single lines, the Amp^r and Tet^r genes allowing selection and the *OR*1 region replication in *E. coli*. Other sequences derived from *S. cerevisiae* DNA are represented by boxed regions, the gene within the hatched box allowing selection for the plasmid on the appropriate yeast auxotrophic mutant, and the dark box sequence allowing plasmid replication in yeast. (a): plasmid YRp7 (Struhl *et al.* 1979) has the yeast *TRP*1 gene and *ARS*1 chromosomal replicator sequence. (b) and (c) are high copy number vectors based on the replication and maintenance sequences of the *S. cerevisiae* 2 μ plasmid (a) replication origin, *OR*1, and an adjacent gene, *REP*3 (not shown), (Jayaram *et al.* 1983). (b): pMA3a (Dobson *et al.* 1982), has *LEU*2 for selection on *leu2*$^-$ yeast and (c): pLG669-Z (Guarente *et al.* 1982) has *URA*3 for selection on *ura3*$^-$ strains. pLG669-Z synthesizes a *CYC*1-*lacZ* fusion protein in yeast, using the *CYC*1 promoter sequence of the *URA*3-*CYC*1 integenic region. By manipulation of these sequences the plasmid can be used to investigate promoter control (Guarente *et al.* 1982).

shown) into which may be inserted DNAs of interest. They contain sequences of bacterial plasmids that allow replication and selection in *E. coli,* in addition to yeast DNA. Their sequences are more easily manipulated using *E. coli* than yeast, and usually it is the final constructs that are introduced into yeast for functional analysis. However, the plasmids can be reisolated from yeast transformants and used to transform *E. coli,* and this is often performed to check that their sequences have not undergone rearrangement in yeast. Plasmid YRp7 (Fig. 3.4(a), Struhl *et al.* 1979) contains the yeast *TRP*1 gene as a selectable marker and *ARS*1, a chromosomal autonomously replicating sequence or replication origin, which allows replication in yeast. YRp7, like other *ARS* plasmids, is only present in a fraction of the cells grown under selective conditions and is lost rapidly in the absence of selection (Kingsman *et al.* 1979). The stability of ARS plasmids can be increased through the addition of a centromere (*CEN*). Although a centromere stabilizes the plasmid in the absence of selection it reduces the copy number to one per cell (Clarke & Carbon 1980). In addition to the *ARS* plasmids, there are shuttle vectors based on the DNA replication and maintenance sequences of the *S. cerevisiae* 2 μm plasmid. Examples of these are shown in Fig. 3.4(b) and (c). The 2 μm plasmid has three genes; *REP*1, *REP*2, and *REP*3 that are essential for its replication and maintenance (Jayaram, *et al.* 1983). For a plasmid bearing the 2 μm DNA replication origin (*OR*1) to replicate in [cir^+] *S. cerevisiae, REP*1 and *REP*2 but not *REP*3 can be provided in *trans* by the 2 μm DNA present. Thus whereas the earlier 2 μm-based shuttle vectors had all of the 2 μm sequences (Beggs 1978), more recently developed ones such as those in Fig. 3.4, have only the *OR*1 region and *REP*3 gene of 2 μm DNA but must be introduced into strains having endogenous 2 μm plasmid [cir^+] in order to replicate. These 2 μm-based shuttle vectors are remarkably stable in the absence of selection, and are present in high copy number. Certain ones, including pMA3a (Fig. 3.4(b)), have a *LEU*2 gene which is not very efficient because it has a truncated promoter and are present at up to 100 copies per cell in [cir^+] strains (Dobson *et al.* 1982, Kingsman *et al.* 1985). The high copy number of these 2 μm-based shuttle vectors can be especially advantageous in studying the expression in yeast of a promoter or gene fusion inserted into one of these vectors. Because this sequence is amplified inside the cell not only are very low levels of its transcription measurable, but there is a good probability that any encoded protein will be overproduced, thereby facilitating detection and purification.

Despite the usefulness and versatility of circular shuttle vectors in analysing the activity of promoter and gene fusions, they do have certain failings which must always be considered in the interpretation of expression studies. Firstly, since they are circular a *cis*-acting element upstream of an ORF could potentially exert its influence right around the plasmid DNA to also act downstream of the ORF. Secondly, their bacterial sequences may not be inert and are often partially transcribed in yeast (Kingsman *et al.* 1985). They may therefore influence expression of the adjacent yeast DNA. Since bacterial sequences are not essential for these plasmids to operate in *S. cerevisiae* it is possible to remove them from plasmid DNA before transfer to this organism.

3.3.3 Integrative vectors

Plasmids which replicate independently of the yeast chromosome through the presence of an *ARS* or 2 μm sequences transform *S. cerevisiae* at high frequency (10^3–10^5 transformants per μg DNA; Struhl 1983). DNAs that lack replication sequences can only transform cells by homologous DNA recombination, and transform much less efficiently (at most 1–10 transformants per μg DNA). However, these integrative DNAs are invaluable, since they enable replacement of a gene with an altered copy of the same sequence without altering the chromosomal environment (Fig. 3.5). This approach to analysis of promoter and gene fusions avoids the aforementioned complications in the interpretation of expression data based upon circular DNA constructs containing bacterial sequences. It is also the most unequivocal way of analysing the function of point mutations, deletions, insertions, and inversions in yeast control elements. Usually only a single copy of the altered gene is inserted into a cell using an integrative vector, although multicopy integrative vectors based on the Ty transposable element are now being developed (Kingsman *et al.* 1985).

The procedure for the precise replacement of a chromosomal gene (X) with an altered copy of the same gene (X′) shown in Fig. 3.5 was originally described by Rothstein (1983). It involves two steps of homologous recombination. These involve linear DNAs with ends that correspond to sequences upstream and downstream of the gene, since double-stranded breaks in DNA increase the frequency of homologous recombination several hundredfold. The first step is to disrupt the existing chromosomal copy of gene X. To do this one can transform, using a linear fragment having a deletion in the X sequence, scoring for loss of gene X′s function (Fig. 3.5)a)). Alternatively one can use a linear fragment of the sequence of X disrupted by a selectable gene such as *TRP*1, selecting in this instance for tryptophan prototrophy after transformation of a *trp*1 strain (Fig. 3.5(b)). The second homologous recombination step involves replacing the inactive copy of gene X (that has either a deletion or insertional inactivation) by an altered copy (X′) constructed *in vitro.* Again the transforming DNA is ideally linear with termini that cause replacement of the resident inactive copy of X. One can select either for restoration of gene activity (Fig. 3.5(a)) or for loss of the inserted gene (*TRP*1^+ to *trp*1^- in Fig. 3.5(b)).

Whether a promoter or gene fusion, or a heterologous gene, is introduced into *S. cerevisiae* on a replicative or integrative vector will depend upon the level of expression and the strain stability that is required. Expression level will be a reflection both of promoter strength and gene dosage or copy number. Gene replacement via integrative vectors has the advantage that the sequences flanking the altered gene are the same chromosomal sequences that flank the wild-type copy of this gene.

3.3.4 Assays for promoter activity

Using promoter and gene fusions (Fig. 3.3) to study sequence function it is necessary both to introduce these sequences into cells upon suitable vectors and have a reliable assay for promoter activity. The levels of the transcripts of a particular sequence can be assayed directly by Northern blot analysis (Meinkoth & Wahl 1984) of either total cellular RNA, if the mRNA is abundant, or RNA enriched for polyadenylated sequences if the mRNA is not abundant. Alternatively an enzyme whose synthesis is under the control of the promoter can be assayed. Levels of expression can be determined by fusing a gene to the ORF of the *lacZ* (β-galactosidase) gene of *E. coli. S. cerevisiae* unlike *Klyveromyces lactis,* has no endogenous β-galactosidase activity so that the functioning of *lacZ* fusions introduced on shuttle

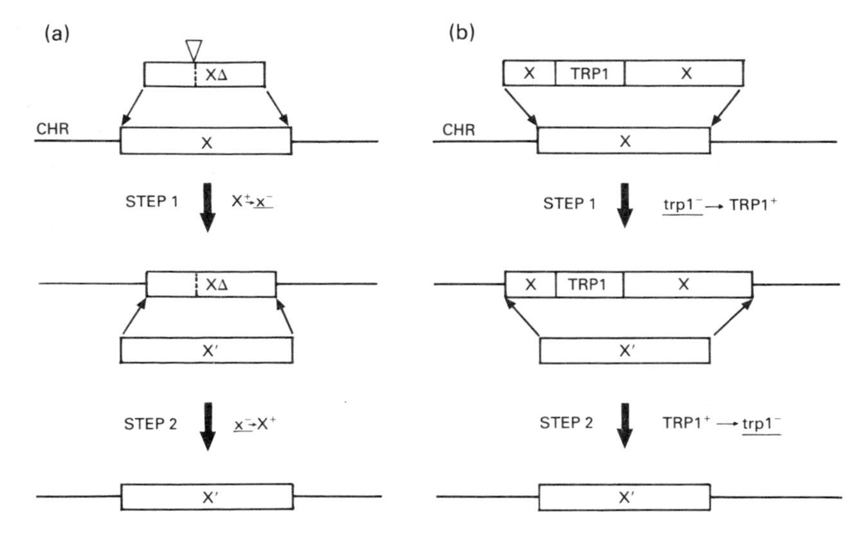

Fig. 3.5 — The use of integrative vectors for the replacement of a gene with an altered copy of the same gene at identical chromosomal locations (Rothstein 1983). Two sequential homologous recombination steps are involved, at the sites indicated by the small arrows. The transforming DNAs are linear with double stranded termini at these points. These termini are highly recombinogenic. In the first step the gene X in chromosomal DNA (CHR) is disrupted by a linear copy of the same sequence that either has a deletion, as in (a), or the insertion of a selectable gene such as *TRP*1, as in (b). In the second step the deleted or *TRP*1-inserted sequence is replaced by a mutant copy of gene X (X′) constructed *in vitro*. Since X′ becomes inserted at the chromosomal location originally occupied by X, any altered expression of X′ can only reflect the sequence alterations introduced *in vitro*.

vectors is readily determined by β-galactosidase assays. In gene fusions involving the ORF of *lacZ* a hybrid protein will be synthesized, provided that the translational reading frame is in register across the junction. Such a hybrid protein will derive its N terminal sequences from the gene under study and its C terminal sequences from the *lacZ* gene. Such fusions are active because the N terminal sequences of *E. coli* β-galactosidase are not essential for enzymatic activity. Vectors have been constructed which contain an intact *lacZ* ORF lacking a promoter, yet preceded by restriction sites (inserted as synthetic oligonucleotide linkers) convenient for insertion of DNA fragments and giving in-frame gene fusions. Insertion of a yeast promoter or N-terminal gene fragment into such vectors gives an expression of β-galactosidase in yeast corresponding to the expression of the yeast sequence. Such vectors are especially useful for the identification of promoter sequences and their regulation (Guarente & Ptashne 1981, Rose *et al.* 1981). A cloning site in these vectors between a promoter and the *lacZ* ORF can also be used for the identification of sequences that constitute sites of transcription termination (Casadaban & Cohen 1980).

An example of a 2μm-based shuttle vector containing a CYC1-*lacZ* gene fusion is shown in Fig. 3.4(c). By substitution of UAS sequences in the promoter region upstream of the CYC1-*lacZ* ORF of this plasmid and assaying the β-galactosidase synthesis of these constructs in yeast, the operation of different yeast UAS elements can be investigated (Guarente *et al.* 1982). The main disadvantage to the use of *lacZ* gene fusions in measurement of promoter activity arises if an efficient promoter is under study and its activity changes transiently in response to a certain cellular stimulus. The activation of the highly efficient phosphoglycerate kinase (PGK) promoter by heat shock of yeast cells is one example of this (Piper *et al.* 1986). The fluctuation of PGK mRNA levels can be readily observed by direct Northern blot analysis of cellular RNAs, but being transient would not be satisfactorily reflected by β-galactosidase assays of a PGK-*lacZ* fusion.

3.4 TRANSCRIPTIONAL PROMOTERS FOR RNA POLYMERASE II

3.4.1 As found in higher eukaryotes

In higher eukaryotes RNA polymerase II initiates transcription 30–40 bp downstream of the TATA-like sequence (Darnell 1982, Breathnach & Chambon 1981). However, the efficiency of the promoter as determined by the frequency of transcription initiation, is determined by other *cis*-acting regulatory sequences. The characteristics of these regulatory elements are being established through their manipulation *in vitro,* and observing their effect in different locations with respect to promoter sequences both *in vivo* and in *in vitro* transcription systems (Manley 1982, Breathnach & Chambon 1981). One class of regulatory elements (the 'upstream elements') acts at close range, being less than 100 bp from the TATA box. They control transcription in response to physiological signals, probably through the interaction of *trans*-acting regulatory proteins with their DNA sequences. The other class of regulatory sequences (the 'enhancers') operate over several kilobases from the TATA box, and can mediate cell-specific or hormone-responsive control over mammalian genes. Their ability to operate over several kilobases from the TATA box, also when inverted, and often when 3′ to the transcription unit, suggests that they may exert their influence by effects on chromatin structure and DNA topology. *Trans*-acting factors are able to interact with enhancers to influence transcription, but may do so via protein–protein rather than protein-DNA interactions (Borrelli *et al.* 1984, Kingston *et al.* 1985).

3.4.2 As found in yeast

Yeast promoters have both a TATA-like sequence and regulatory sequences that can be positioned within several hundred nucleotides of the transcription start. They are therefore less similar to bacterial promoters, where practically all of the promoter and its control sequences lie within 100 nt of the transcription startpoint, than to mammalian promoters. The TATA box is an important part of certain *S. cerevisiae* promoters, since its deletion severely reduces transcription levels without affecting regulation (Struhl 1982, Guarente & Mason 1983). In general, though, it is found further upstream of the startpoint for transcription than in genes of higher eukaryotes. Some yeast promoters have multiple TATA boxes. For example the *CYC*1 gene has TATA regions at −154, −106, −52, and −22 nt relative to the first transcriptional initiation site (Smith *et al.* 1979). Probably owing to this diversity of TATA sequences, downstream initiation upon the *CYC*1 gene occurs in six major regions at: +1, +10, +16, +25, +34, and +43 nt to give leader sequences on *CYC*1 mRNA of very heterogeneous length (Faye *et al.* 1981). *ADH*1 and *SUC*2 also have two transcription start sites (Bennetzen & Hall 1982, Carlson & Botstein 1982). In the case of *SUC*2 the choice of transcription start site reflects glucose regulation at this locus (section 3.5.1). Although yeast TATA elements are further from transcription startpoints than in higher eukaryote genes, deletions changing the spacing between the TATA sequences and the region of transcription initiation frequently do not change these start sites (McNeil & Smith 1985). Thus at least part of the information determining mRNA initiation sites is within the DNA sequence at these sites. Analysis of promoters suggests that an A with a short region of purines on each side acts as a preferred site of initiation (McNeil & Smith 1985).

UAS elements that lie upstream of the TATA box and which activate transcription in response to particular physiological signals appear to be commonplace in yeast promoters (Guarente 1984). They are discrete DNA sequences that often lie some hundreds of nucleotides upstream of the initiation region. This, and the variability allowed in their spacing from transcription initiation sites, distinguishes them from the activation sites found on prokaryotic genes. The functioning of particularly well characterized UAS sites is discussed in sections 3.5 and 3.6. Many yeast promoters can be expected to have different UAS sites, overlapping or arranged in tandem, by which they respond to different physiological stimuli. An example of this is the *CYC*1 gene (section 3.5.4). In this manner a single locus can be under complex regulatory control, yet its transcription can always initiate at the position specified by the TATA box. This mechanism of gene control has parallels to that used in higher eukaryotes, yet contrasts strongly with the situation in bacteria where genes under complex control often have multiple differently-regulated promoters which direct transcription initiation from multiple sites (see for example DiLauro *et al.* 1979, Sancar *et al.* 1982, Johnson *et al.* 1983).

Elegant genetic studies have shown that at least some UAS regions comprise the binding sites on DNA for regulatory gene products, presumably proteins. This binding activates transcription of the adjacent promoter (sections 3.5 and 3.6). It has yet to be established whether these *trans*-acting factors also interact directly with RNA polymerase II or indirectly assist polymerase entry to the promoter by altering chromatin structure and DNA topology. There is recent evidence that the UAS in the *GAL*1-*GAL*IO intergenic region (Fig. 3.6) is the preferred site in this region for interaction with purified yeast RNA polymerase II, and that this interaction produces a conformational change detected as torsional strain in the DNA (DiMauro *et al.* 1985). Thus this UAS may act as an entry site for RNA polymerase II prior to formation of the open promoter complex. Consistent with this

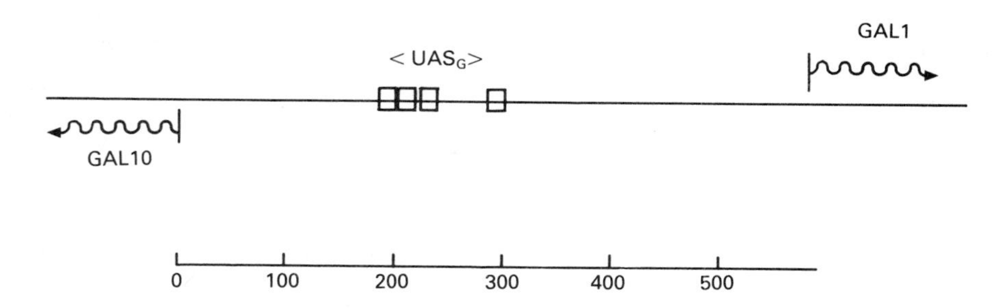

Fig. 3.6 — The portion of yeast genome containing the *GAL*1 and *GAL*10 promoters. The divergent transcripts (St John & Davis 1981) are indicated by wavy lines. The 17 bp dyad symmetry elements that bind *GAL*4 protein during transcription activation (Gininger *et al.* 1985) are shown as boxes. The lower scale indicates distance in nucleotides.

notion are the findings that the UAS sites of *CYC*1 and *GAL*1-*GAL*10 are both inoperative when positioned in either orientation downstream of the TATA boxes of these genes (Guarente & Hoar 1984, Struhl 1984).

The protein factors that interact with UAS regions are often, as underlies their function, themselves the target of complex controls (e.g. *GAL*4 and *GCN*4, section 3.5). Most are probably nuclear proteins, since some have in their amino acid sequences both a DNA binding site and a distinctive protein region that causes them to enter and remain in the nucleus (Silver *et al.* 1984, Hall *et al.* 1984). Many of these factors (e.g. *GCN*4, *HAP*2, *HAP*3, *MATα*1 and *STE*12) are activators of several genes.

One example has been found of an activator sequence within the ORF of a *S. cerevisiae* gene (Kingsman *et al.* 1985). The phosphoglycerate kinase (*PGK*) gene has such a sequence, and it is essential for efficient *PGK* promoter activity. A detailed account of the *PGK* and *TRP*1 promoter regions is given in Kingsman *et al.* (1985).

There is now mounting evidence that yeast genes can be not only activated but also repressed through protein binding to their upstream regions. Just as activation depends on the binding of proteins to UAS regions, repression of transcription depends on proteins interacting with sites on the DNA other than UASs (Brent 1985). It is still not clear how such protein binding to DNA represses transcription. It might interfere sterically with the binding of positive regulatory proteins or RNA polymerase to UAS or TATA regions, or it might prevent transmission of the UAS signal to the TATA region.

That yeast gene repression can occur by protein binding was shown using a model system in which *S. cerevisiae* cells were engineered to produce the *E. coli lexA* protein. When the *lexA* operator sequence was placed into the same cells between the UAS and TATA box of the *GAL*1 promoter, the *lexA* protein bound to its cognate operator and repressed *GAL*1 transcription 3 to 10-fold. the positioning of the *lexA* operator in the construct influenced the degree of repression obtained, and repression was only observed when the *lexA* operator was downstream of the UAS (Brent & Ptashne 1984).

Certain yeast genes, such as *PGK* (Kingsman *et al. 1985) and CYC7* (Wright & Zitomer 1985), have been shown to have a DNA region that exerts a negative effect upon their expression. However, the best evidence for protein repressors in yeast, and repression sites on upstream regions of genes comes from the *MAT*α2 and *al*-α2 repressions of cell type control (section 3.6.1). Expression of the Ty transposable element sequences is also subject to negative regulation by the protein products of the *SPT* genes, of which their are at least seven (Roeder *et al.* 1985). A similar situation of multiple negative regulators of expression of a specific sequence is found in the *SIR* loci (section 3.6.2).

3.5 INDUCTION OF *S. CEREVISIAE* GENES BY POSITIVE REGULATION OF TRANSCRIPTION

3.5.1 Derepression of glucose repressible genes

Classical genetics has not yet uncovered a simple repressor–operator control system for the catabolic enzymes of yeast of the kind so common in bacteria. By introducing into *S. cerevisiae* both an *E. coli* repressor gene and a yeast gene into which has been inserted on *E. coli* operator sequence (section 3.4.2) it has been shown that such a system could operate. Many *S. cerevisiae* genes for fermentative respiration are activated by glucose (McAlister & Holland 1982), but the mechanism of this activation has yet to be established. Other genes are repressed by glucose (Entian & Frohlich 1984). Their derepression in media such as glycerol occurs at the level of transcription, not through the removal of a repressor, but through the activation of a *trans*-acting positive regulatory factor which then binds to a cognate UAS element. So-called 'glucose repression' has therefore to be considered as the inhibition of a UAS-mediated positive gene activation process rather than negative regulation.

One of the two genes for yeast hexokinase, *HXK*2 encoding the PII isoenzyme, has been implicated in the mechanism of glucose repression. Some *hxk*2 mutants display defective glucose repression even though their hexokinase PII isoenzyme is catalytically active (Entian & Frohlich 1984, Entian *et al.* 1984, 1985).

The *ADH*2 (or *ADR*2) gene for alcohol dehydrogenase is but one of several glucose-repressed genes of *S. cerevisiae.* Many mutants of *ADH*2 fail to derepress when glucose is removed, and many of these have Ty transposable element insertions in the vicinity of the *ADH*2 TATA box region (Russell *et al.* 1983). Another class give greater *ADH*2 derepression than in wild-type cells, and are associated with increase in the length of a poly dA, poly dT tract 80 bp upstream of the TATA box. Evidence that the DNA region 200–1000 nt upstream of the transcription start of *ADH*2 confers glucose-regulated expression was obtained by inserting this sequence upstream of *ADC*1; the gene for the constitutive alcohol dehydrogenase. *ADC*1 then became glucose regulated (Beier & Young 1982,

Russell *et al.* 1983). The *ADR*1 gene encodes one *trans*-acting positive regulator of *ADH*2, and its ability to bring about 100-fold *ADH*2 derepression requires a 22 bp region of perfect dyad summetry at −215 nt with respect to the *ADH*2 transcription start site. If this dyad is removed, derepression in the absence of glucose is tenfold less efficient and *ADR*1 independent. Evidence for a positive activation of *ADH*2 and absence of a repressor is suggested by *in vivo* competition experiments. Cells containing multiple copies of the *ADR*1 gene on a high copy number plasmid show constitutive *ADH*2 expression on glucose and superinduction after derepression. Also multiple copies of the *ADH*2 promoter do not relieve repression but inhibit transcription of the chromosomal *ADH*2 allele during derepression. This inhibition requires the 22 bp dyad element and is not overcome by excess *ADR*1 product. *ADR*1 itself is not glucose regulated, and it is probably not the only activator of *ADH*2 transcription.

The locus *SUC*2 encodes two differently-regulated forms of invertase. A *SUC*2 mRNA synthesized constitutively at low-level encodes an intracellular invertase, whereas another *SUC*2 mRNA 100 to 1000-fold repressed by glucose encodes a secreted invertase (Carlson & Botstein 1982). The glucose-repressed mRNA is longer, reflecting usage of an earlier transcription startpoint on the gene. Deletion analysis of the upstream noncoding region of *SUC*2 has identified a sequence: −650 to −418 nt from the transcription start which is needed for derepression of secreted invertase. This sequence can confer glucose-regulated expression to other genes, and contains the heptanucleotide sequence A(A/G) GAAAT repeated four times (Sarokin & Carlson 1985). The products of genes designated *SNF*1-*SNF*6 are involved in derepression of *SUC*2 and other glucose repressible genes (Carlson, *et al.* 1981). Their relationship to *HXK*2 and *ADR*1 controls has yet to be reported.

3.5.2 Induction of structural genes of the galactose–melibiose regulon

The transcription of the genes that encode enzymes of galactose utilization in *S. cerevisiae* (*GAL*1, *GAL*7 and *GAL10*) is induced more than 5000-fold by galactose (Hopper *et al.* 1978, St John & Davis 1981). This system has attracted a lot of attention because it does not involve duplicated genes for a particular function unlike those involved in *S. cerevisiae* maltose and sucrose utilisation. *GAL*1 and *GAL*10 are adjacent genes on chromosome II transcribed in opposite directions from sites approximately 580 bp apart (Fig. 3.6). So efficient is the galactose control over their promoters that their control sequences are now often used in gene constructs to bring sequences under galactose regulation.

The induction of *GAL*1 and *GAL*10 depends on a *cis*-acting UAS element (UAS_G) found midway between these divergently-transcribed genes. UAS_G when placed in front of the TATA box of the *CYC*1 gene was sufficient to render the resultant hybrid promoter inducible with galactose (Guarente *et al.* 1982).

Transcription of *GAL*1, *GAL*7, and *GAL*10 is uninducible in cells having recessive mutations in the *GAL4* gene (Oshima 1982). The observation that these *gal4*$^{-}$ mutations are epistatic to other galactose regulatory mutations was the first indication that the *GAL4* product may be the transcription factor that interacts with UAS_G in galactose induction (Guarente *et al.* 1982, Johnston & Hopper 1982). *GAL4* protein has a sequence that causes it to enter and remain in the nucleus (Silver *et al.* 1984). It is inhibited in its function by the *GAL80* product (Oshima 1982), yet the *GAL4* positive regulatory element is epistatic to this negative *GAL80* element. Experiments determining the protection against methylation of DNA due to protein binding have revealed that when *GAL4* binds to UAS_G *in vivo* it interacts with four related dyad-symmetric sequences (Fig. 3.6) each of 17 bp (Giniger *et al.* 1985). Only one of these sequences is sufficient to confer nearly wild-type levels of *GAL4*-mediated inducibility to the *GAL1* or *CYC1* genes, and the fact that four copies are found in UAS_G suggests that multimeric forms of *GAL4* protein may be binding to this UAS. DNA footprinting experiments also show that *GAL80* mediated repression does not interfere with *GAL4* protein binding UAS_G (Gininger *et al.* 1985). The inducer, galactose, interacts with this *GAL80* protein to prevent its functioning (Nogi *et al.* 1977, Perlman & Hopper, 1979).

GAL4 also promotes transcription of *MEL1*, a gene that encodes the melobiose-cleaving enzyme α-galactosidase (Summer-Smith *et al.* 1976). Expression of *MEL1* is induced about 100-fold by galactose. However, the domain of *GAL4* required for transcription of *GAL1*, *GAL7*, and *GAL10* differs from the domain that promotes transcription of *MEL1*. Loss of the 3′ terminal aminoacids of *GAL4* by deletion or use of nonsense mutants abolishes the former but not the latter function (Matsumoto *et al.* 1980). The *GAL4* locus actually encodes two different mRNAs and proteins (Laughon *et al.* 1984). The smaller protein (92 kd) lacks the N-terminal 78 aminoacids of the larger protein (100 kd). Frameshift mutations in the coding region unique to the larger protein have shown that this species is essential for complementation of *gal4*$^{-}$ mutants and for *GAL80* operation. Also site-directed mutagenesis of the AUG initiation codon of the smaller *GAL4* protein has been used to show that the larger *GAL4* protein is alone necessary and sufficient for UAS_G transcriptional activation (Laughon *et al.* 1984).

Glucose strongly represses the galactose regulon, in a process that is independent of *GAL80* (Matsumoto *et al.* 1981). DNA footprinting has shown that glucose repression of the *GAL* genes is accompanied by loss of the *GAL4* protein footprint on UAS_G (Giniger *et al.* 1985). The glucose might inhibit the *GAL4*-UAS_G interaction directly, or affect this interaction indirectly through another protein–DNA interaction close to, or overlapping, UAS_G.

3.5.3 Induction of structural genes of the maltose regulon

There appear to be two pathways for maltose utilization in yeast: a fermentative pathway requiring any one of five unlinked *MAL loci (MAL1*-4 and *MAL6*) and independent of mitochondria, and an oxidative pathway for which mitochondrial functions are indispensable. The fermentative genes are coordinately induced by maltose and repressed by glucose at the transcriptional level. The maltase (*MAL*62) and maltose permease (*MAL*61) genes of the *MAL*6 locus of *S. carlsbergensis* are, like *GAL*1 and *GAL10* (Fig. 3.6), transcribed divergently. Their intergenic region is thought to act in coordinate regulation in a similar way to the region that separates *GAL*1 and *GAL*10 (Hong *et al.* 1984, Cohen *et al.* 1985). The same *MAL*6 locus is also thought to encode a positive reglatory protein (*MAL*63 product) required for maltose induction, although some mutants constitutive for maltase and maltose permease are known to map outside of the *MAL*63 region.

3.5.4 Induction of genes for iso-cytochromes c and haem biosynthesis

The gene for iso-1-cytochrome c (*CYC*1) has two homologous sequences located in tandem upstream of the TATA region that mediate transcriptional activation by haem. These UAS sites are at −275 nt (UAS1) and −225 (UAS2) relative to the transcription startpoint (Guarente 1984). *CYC*1 is uninduced under anaerobic conditions and conditions of haem deficiency, UAS1 is partly induced in the presence of haem in glucose media, and both UAS1 and UAS2 are fully active in the presence of haem in nonfermentable media. Both UAS sites function when their orientation with respect to the TATA box is inverted, but not when present downstream of the TATA box (Guarente & Hoar 1984, Guarente *et al.* 1984). Either UAS1 or UAS2 inserted upstream of the TATA box of the *LEU*2 gene is sufficient to make *LEU*2 haem activatable (Guarente *et al.* 1984). The induction of *CYC*1 by haem is reduced by recessive mutations at the regulatory loci *HAP*1, *HAP*2, and *HAP*3. These *HAP* genes are thought to encode activator proteins which are either nonfunctional or not synthesized in the absence of haem (Pinkham & Guarente 1985). Mutations at *HAP*1 affect the operation of UAS1 but not UAS2, whereas mutations in *HAP*2 or *HAP*3 affect UAS2 only. The two haem-responsive UAS sites of *CYC*1 seem therefore to be regulated by separate *trans*-acting factors (Guarente *et al.* 1984). The *hap*2.1 and *hap*3.1 mutants are pleiotropic nuclear *petites* affecting not only *CYC1* but also other cytochrome and respiratory genes. *HAP*2 and *HAP*3 are therefore part of a general activation system for respiratory functions (Guarente *et al.* 1984, Pinkham & Guarente 1985). *HAP*2 gene transcription is suppressed fivefold by glucose (Pinkham & Guarente 1985).

Mutations of UAS1 of *CYC*1 indicate that the *HAP*1 responsive sequence is a pentamer, variants of which are repeated six times over a 68 bp region. Mutations affecting only one of the repeats or their spacing reduces activity 2–10 fold (Lalonde, *et al.* 1986). Transcription of *CYC*7, the *S. cerevisiae* gene for iso-2-cytochrome c, occurs at only about 5% of the level of transcription of *CYC*1. *CYC*7 is also regulated by *HAP*1, but its transcription is less dependent on intracellular haem than the transcription of *CYC*1. *CYC*7 has a *HAP*1-responsive UAS with four repeats of the pentamer sequence found in UAS1 of the *CYC*1 gene, that are distributed over 120 bp and possibly overlap a repression site (Wright & Zitomer 1985, Iborra *et al.* 1985).

The *HEM*1 gene, encoding S-aminolevulinate synthetase the first enzyme of haem biosynthesis, is under positive control by *HAP*2 and *HAP*3. This *HAP*2/*HAP*3 activation of *HEM*1 is inhibited by haem; whereas the activation of cytochrome genes by the same loci is haem-requiring (Keng & Guarente 1986). See also Chapter 5, pp. 138–139.

3.5.5 Induction of genes for aminoacid biosynthetic pathway enzymes

Starvation of yeast cells for a single aminoacid stimulates the transcription of at least 30 genes for the enzymes of different amino acid biosynthetic pathways. This coordinated derepression involves both *cis*-acting and *trans*-acting elements, and is often referred to as the 'general control system' (reviewed by Hinnebusch 1985). Derepression is eliminated by mutations in several loci for positive (*GCN*) and negative (*GCD*) *trans*-acting effectors. Mutants that are gcn^- are nonderepressible, while those that are gcd^- are constitutively derepressed. Several temperature-sensitive cell cycle (*cdc*) mutants are also derepressed at the nonpermissive temperature and can therefore be classified *gcd*. $gcn4^-$ mutations are epistatic to all other mutations affecting general control, indicating that *GCN*4 (formerly called *AAS*3) encodes the proximal positive effector of the system (Hinnebusch & Fink 1983). *GCN*4 expression is itself subject to general control (Hall *et al.* 1984, Thireos *et al.* 1984), being negatively controlled by *GCD*1, which is in turn repressed by *GCN*1, *GCN*2, and *GCN*3. *GCN*4 mRNA has an unusually long 5′ leader sequence (600 nt) which, because it contains four small ORFs, may be controlled at the translational level (Thireos *et al.* 1984).

The upstream regions of genes subject to general control contain copies of a short sequence that mediates *GCN*4 action. It has the consensus sequence: TGACTC, followed by a downstream dT tract. Several copies of this sequence, and variants thereof, are present in the upstream regions of *HIS*3 and *HIS*4 (Struhl 1982, Donahue *et al.* 1983, Hinnebusch & Fink 1983; Zalkin & Yanofsky 1982). Also a synthetic *HIS*4 regulatory element containing at least one TGACTC can confer general control upon the *CYC*1 gene (Hinnebusch *et al.* 1985). The TGACTC sequence comprises part of the binding site for the *GCN*4 product. *GCN*4 protein synthesized *in vitro* specifically protects a 20 bp region of the *HIS*3 promoter that includes the TGACTC sequence proximal to the TATA box from deoxyribonuclease 1 digestion (Hope & Struhl 1985).

3.5.6 Induction of phosphatase genes

Acid phosphatase provides a convenient analysis system for regulatory functions since it is located in the periplasmic space of the cell envelope, and its activity can be monitored by staining colonies by the diazo-coupling method (Schurr & Yagil 1971). Colonies expressing acid phosphatase are red, while those lacking activity are white. Gene fusions with the acid phosphatase ORF therefore provide a simple system to study promoter element function.

Two of the *S. cerevisiae* acid phosphatase genes, *PHO*3 and *PHO*5, are found adjacent to each other on chromosome II. *PHO*3 is expressed constitutively. *PHO*5, and a third unlinked acid phosphatase gene, are repressed in media containing inorganic phosphate but transcriptionally induced more than 50-fold in response to the absence of phosphate (Bajwa *et al.* 1984). *PHO*5, like the galactose-inducible genes (section 3.5.2), is controlled by several unlinked positive (*PHO*4, *PHO*2, *PHO*81) and negative (*PHO*80, *PHO*85) acting regulatory genes (reviewed by Oshima 1982).

Alkaline phosphatase is located wholly inside the yeast envelope, but can be detected by permeabilization of cells to substrate. *PHO*8, the gene for repressible alkaline phosphatase, requires some of the same genes for full transcriptional derepression in response to medium phosphate levels that are needed for *PHO*5 activity (Oshima 1982). In the repressed state basal *PHO*8 activity is much greater than with *PHO*5 encoded acid phosphatase, and *PHO*8 derepression requires at least one gene (*PHO*9) not needed for *PHO*5 derepression (Kaneko *et al.* 1985). *Neurospora crassa* has a similar regulatory system to *S. cerevisiae* for the control of acid and alkaline phosphatase synthesis (reviewed in Oshima 1982).

3.6 GENE ACTIVATION AND REPRESSION IN THE DETERMINATION OF MATING TYPE

3.6.1 Controls mediated by the mating type loci and STE genes

Much of our understanding of how *S. cerevisiae* activates and represses genes has been derived from the extensive molecular genetics of cell type determinants (reviewed by Herskowitz & Oshima 1981; Sprague, Blair & Thorner, 1983). Haploid cells of mating type *a* mate efficiently with haploid cells of mating type α to form a third cell type, the *a*/α diploid cell. Each of these types of cell produces transcripts from a distinct set of genes that govern its mating or sporulation ability. As a result only α cells secrete the α-factor mating pheromone and make both surface agglutinin specific for *a* cells and the receptor for *a*-factor pheromone. Also only *a* cells secrete *a* factor, agglutinate with α, and synthesize the α-factor receptor. Diploid *a*/α cells do not produce or respond to pheromones, do not agglutinate, and are able to sporulate in response to the appropriate physiological stimuli.

The mating type of haploid cells is determined by the allele present at the mating type locus on chromosome III (Fig. 3.7). *S. cerevisiae a* cells have the *MATa* allele, α cells the *MAT*α allele, and diploid *a*/α cells are heterozygous at their MAT loci. The *MAT* locus encodes *trans*-acting factors that directly control unlinked cell-type specific functions. On the basis of mutations in mating type genes and structural analysis of *MATa* and *MAT*α, Nasmyth *et al.* (1981), Tatchell *et al.* (1981), and Strathern *et al.* (1981) proposed that the *MAT* alleles encode three regulatory activities α1, α2, and *a*1.α2. Their model for the role of these proteins, modified to include the function of *STE*12 (Fields & Herskowitz 1985) is shown in Fig. 3.8. *MAT*α encodes two products: MATα1 which acts with *STE*12 to activate unlinked α-specific genes, and *MAT*α2 which is a negative regulator of unlinked *a*-specific genes. *MAT*α1 and *STE*12 are not required for each other's synthesis,

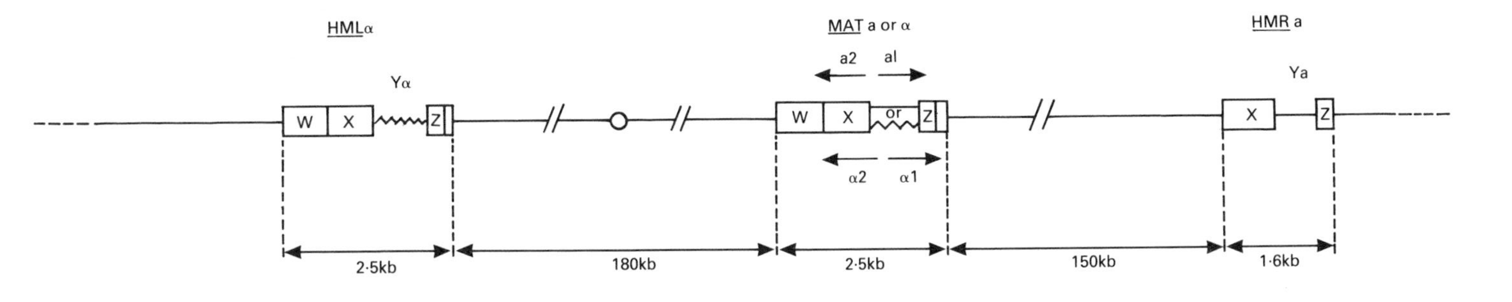

Fig. 3.7 — The three Mating Type Loci on Chromosome III of Yeast. The straight line represents chromosome III, the boxes represent the mating type loci, and the circle represents the centromere, *CEN*3. Regions X and Z are present at all three loci, whereas W and a small region adjacent to Z are found at *HML* and *MAT*, but not *HMR*. Ya is a 642 bp *a*-specific sequence, and Yα is a 747 bp α-specific sequence. The *MAT a*1, *a*2, or α1, α2 genes are also present at HMR and HML respectively, but are repressed at these loci by the products of the five SIR genes. During mating-type interconversion, the sequences at MAT (*a* or α) are replaced by α or *a* sequences copied from one of the two HM loci. (Reprinted from Brand *et al.* (1985) by permission of *Cell*, copyright © 1985 *Cell*).

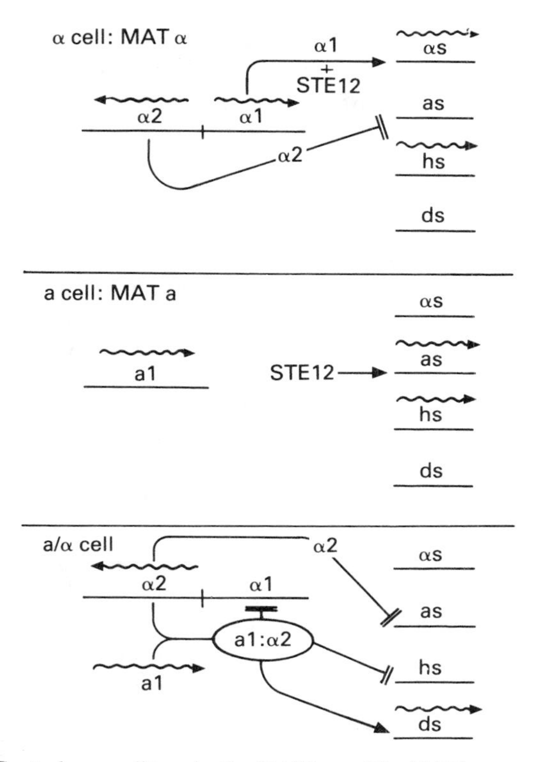

Fig. 3.8 — Control over cell type by the *MAT* locus. The *MAT* locus gene expression is shown on the left, and the expression of unlinked genes controlled by this *MAT* expression is shown on the right (αs = α-specific genes, *as* = *a*-specific genes, *hs* = haploid – specific genes, *ds* = diploid specific genes). A wavy line indicates gene expression; a line with an arrowhead indicates activation of expression, and a double bar indicates repression.

This diagram is a compilation of the results of studies described by Miller *et al.* (1985, Ammerer *et al.* (1985), and Fields & Herskowitz (1985). In an α cell the *MAT*α1 product stimulates expression of αs genes and *MAT*α2 inhibits expression of *as* genes. In an *a* cell *MAT*α1 product is absent and αs genes are not therefore expressed, and the absence of *MAT*α2 leads to the expression of *as* genes. Both *a* and α cells express *hs* genes, but in *a*/α cells *hs* genes are repressed by the presence of both the *MAT*α1 and *MAT*α2 products (the *a*1:α2 complex). Also in *a*/α cells *MAT*α2 inhibits expression of *as* genes and *a*1:α2 inhibits expression of *MAT*α1. Where a role for *STE*12 is shown in activation other *STE* genes may also be involved (Fields & Herskowitz 1985). Also the involvement of *a*1:α2 in the expression of those *ds* genes involved in sporulation must involve factors involved in triggering onset of sporulation.

MATa codes for *MATa*1 only. *MATa*1 is not needed for the *STE12*-requiring *a*-specific gene expression, but acts instead together with *MAT*α2 in *a*/α diploids to suppress haploid-specific genes. The *MAT*α2 repression of *a*-specific genes also operates in *a*/α diploids. For three haploid specific genes, *MAT*α1, HO, and *STE*3, a1-α2 repression has been shown to be at the level of RNA production (Nasmyth *et al.* 1981, Nasmyth *et al.* 1985, Sprague *et al.* 1983).

All the regulation in Fig. 3.8 is thought to occur by control of transcription. The UAS sequences involved in gene activation by the *MAT*α1 and

*STE*12 products, and their properties, have yet to be reported. *STE* genes other than *STE*12 are probably also involved in this activation (Fields & Herskowitz 1985). If *MAT*α1 is produced via a hybrid gene in *a* or *a*/α cells, α-specific genes are produced in the *a* cells but not the *a*/α (Ammerer *et al.* 1985). An additional regulatory system must therefore prevent α gene expression *a*/α cells even with *MAT*α1 present. Haploid-specific genes are defined by their lack of expression in *a*/α diploids, and each might have its own activator. Both *MATa*1 and *MAT*α1 gene products are essential for sporulation, and *a*1-α2 is thought to both repress haploid-specific genes and be required in the activation of sporulation specific genes (dsg, Fig. 3.8). Activation of the sporulation-specific glucoamylase (*STA*) gene by *a*1-α2 is at the level of RNA production, but whether *a*1-α2 actually binds to upstream regions of this gene is not yet clear (Yamashita & Fukui 1985). More research is needed to establish how much of the temporal appearance of sporulation-specific transcripts in *S. cerevisiae* sporulation (Kaback & Feldberg 1985) is *a*1-α2 dependent, and how synthesis of these transcripts is repressed in haploid cells.

*MAT*α1 and *MAT*α2 have sequences reminiscent of the protein-binding domains of bacterial repressor proteins, which suggests that they bind to DNA. Miller *et al.* (1985) initially identified the DNA sequences that are recognized by the *MAT*α2 and *al*-α2 repressions from sequence comparison of upstream regions of genes repressed by these systems. The recognition sequences of the two repression systems share certain sequence elements, presumably since *MAT*α2 protein is involved in both, and yet they are not identical. The sequence (T/C)C(A/G) TGTNN (A/T) NANNTACATCA is sufficient to confer *al*-α2 repression, and the related sequence GCATG-TAATTACCCAAAAAGGAAATTTACATGG can confer *MAT*α2 repression. Both carry the same: ATGT . . . ACAT motif (underlined) in inverted orientation, but space the two halves of this element differently. Miller *et al.* (1985) propose that *MAT*α2 binds to its recognition sequence as a dimer, whereas *a*1-α2 binds as a single *a*1-α2 complex.

Proof that these sequences are alone sufficient to confer repression was obtained by demonstrating the correct repression pattern after their insertion into a promoter (*CYC*1) not normally under the control of the *MAT* locus (Miller *et al.* 1985). In another study aimed at identifying the DNA sequences recognized by *MAT*α2 Johnson & Herskowitz (1985) fused the *MAT*α2 gene to the *E. coli lacZ* ORF. The *MAT*α2-*lacZ* fusion protein made by yeast cells containing this construct represses their *a*-specific genes. Also this fusion protein contains a DNA-binding domain that is able to protect a 33 bp portion of the upstream region of one *a*-specific gene (*STE*6) from deoxyribonuclease 1 digestion. This protected sequence lies between the UAS and TATA box of *STE*6, possibly thereby preventing UAS action when *MAT*α2 is bound. It is equivalent to the sequence identified by Miller *et al.* (1985), and when inserted into the *CYC*1 promoter can also bring *CYC*1 under *MAT*α2 control (Johnson & Herskowitz 1985).

The location of either a *MAT*α2 or a *a*1-α2 repression sequence within a gene is important for its operation, and affects dramatically the degree of repression. The *MATα2* repression sequence is most effective (100–200 fold repression) when inserted between the haem-UAS and TATA box of the *CYC1* gene, and much less effective (less than 20-fold repression) further upstream (Miller *et al.* 1985, Johnston & Herskowitz, 1985). The *MATa1* gene is subject to *MATα2* regulation at a site between its UAS and TATA regions (Siliciano & Tatchell 1984). Also, the sequences upon haploid-specific genes that confer *a*1-α2 repression are more efficient in their action when between a UAS and a TATA region than when downstream of a TATA. The *HO* gene has multiple copies of such a sequence in its 5′ flanking DNA, and although it will bind *MATα2* protein *in vitro*, repression *in vivo* also requires *MATa1* (Johnson & Herskowitz 1985, Nasmyth 1985). These repression sequences are inoperative when inserted into the protein coding regions of genes (Siliciano & Tatchell 1984, Miller *et al.* 1985), and possibly block in some way the transmission of the UAS signal to the TATA.

3.6.2 Repression of the *HMLa* and *HMR* silent copies of mating information by the *SIR* loci

In addition to the mating type information at the *MAT* expression locus, which determines cell type (Fig. 3.8), each cell has silent copies of the *a* mating information (*HMLa*) and α mating information (*HMR*α) located some distance away from *MAT* on chromosome III (Fig. 3.7). At MAT is inserted either a part of the *HMLa* DNA region or a part of the *HML*α DNA region, and it is this which is expressed to determine cell type. On occasion an *a* cell switches to α mating type, or an α cell switches to *a* in a process termed mating type interconversion. When this occurs the mating information at the *MAT* locus is excised and is replaced by a new copy derived from either *HMR*α or *HMLa*. The original silent copy of the locus concerned remains at the inactive site as another copy is inserted at the *MAT* locus in a process that must entail site-specific DNA recombination and DNA synthesis. Also, since the *MAT* locus already contains a DNA segment from either *HMLa* or *HMR*α, the entry of a new unit of DNA into *MAT* must be accompanied by the removal of the resident unit. Mating type interconversion only occurs if the cell is expressing the haploid-specific gene *HO* (repressed in diploids by *a*1-α2 (Nasmyth 1985)). Haploid *HO* cells are heterothallic, and switch between the *a* and α mating types usually within a single generation. Such haploid strains are not stable since they give both *a* and α progeny that mate to give *a*/α diploids. In contrast ho^- haploid strains are homothallic and stable.

The two *HMLa* and *HMR*α silent storage copies of mating information on chromosome III comprise complete copies of the *a* and α genes with their promoters that are not transcribed until moved to the *MAT* locus (Fig. 3.7). Their repression operates by a mechanism specific for these loci and depends on the products of at least 5 unlinked genes (variously called the *SIR* or *MAR* loci). Mutations in any of these genes allow both *HMLa* and *HMR*α to be transcribed and cause sterility. The sites where the products of these *SIR* genes interact with DNA to suppress *HMLa* and *HMR*α have been termed 'silencer' sites since they appear to have functionally opposite effects to the enhancers of eukaryotic cells (Brand *et al.* 1985, Abraham *et al.* 1984). The silencer sequence is located 1.7 kb upstream of *HMLa* and 0.9 kb upstream of *HMR*α, and this repression is therefore unlike that mediated by *MAT*α2 or *a*1-α2 in that its operation is not limited to the confines of promoter sequences.

Although it operates to suppress rather than enhance transcription this silencer element shares several features displayed by the enhancers of higher organisms. Thus it can act to repress a promoter from which it is separated by several kilobases, can operate on two promoters at once, and is able to function upstream or downstream of a gene to effect at least a 100-fold repression. The silencer sequence contains an ARS, not yet identified as a feature of enhancers, that may be a replication origin since it confers upon plasmids the ability to replicate autonomously in yeast. Repression may therefore result from the creation of a structure in chromatin that prevents transcription of adjacent DNA (Brandt *et al.* 1985). In support of this is the finding that the *SIR* regulatory system can repress non-mating type genes such as *TRP*1 and *LEU*2 when these are inserted at the *HMLa* and *HMR*α loci. Also, DNA replication may be important for *SIR* repression to operate since passage through the S phase of the cell cycle is required to achieve *SIR* repression of *HMR*α (Miller & Nasmyth 1984).

3.7 CONCLUSIONS AND FUTURE PROSPECTS

The regulatory mechanisms for enzyme synthesis described in section 3.5 have the advantage that they are controlled by simple effectors such as glucose, galactose, phosphate, or haem. They do not involve programmed control mechanisms as in the yeast cell cycle, or complex cycles of enzyme activity modulation by covalent modification as often found in mammalian cells. Through the action of simple-effector controls the levels of specific enzymes can be increased or decreased by factors of more than 50-fold, and understanding of how these controls operate is advancing rapidly.

As yet we do not have a clear picture of the *cis*- and *trans*-acting controls whereby selective gene expression is achieved during the cell cycle and the stress responses of yeast cells. Examples of sequences expressed at particular phases of the cell cycle are the genes for histones (Hereford *et al.* 1980), certain *CDC* genes (Peterson *et al.* 1985) and the HO endonuclease of mating type switching (Nasmyth 1985a). Nasmyth (1985b) has given preliminary evidence for the sequences on the HO gene that might control its cell-cycle-dependent expression. Examples of stress response genes are those activated in response to DNA-damaging agents such as ultraviolet and ionizing radiations (Kaback & Feldberg 1985, Rolfe *et al.* 1985; also Kelly, this volume), and genes activated in response to heat shock and anoxia (Piper *et al.* 1986). Studies over the next few years will lead to identification of the genetic elements that control expression of cell-cycle and stress-responsive genes, and elucidation of their modes of action. In bacteria stress genes often respond to more than one stress condition (Walker 1984, Neidhardt *et al.* 1984), and the same may also apply in yeast. Thus DNA-damaging agents can cause stimulation of genes for heat shock proteins in *E. coli.* An example of a stress response in *S. cerevisiae* (heat shock) is shown in Fig. 3.9. When cells of *S. cerevisiae* are shifted rapidly from 25°C to 42°C

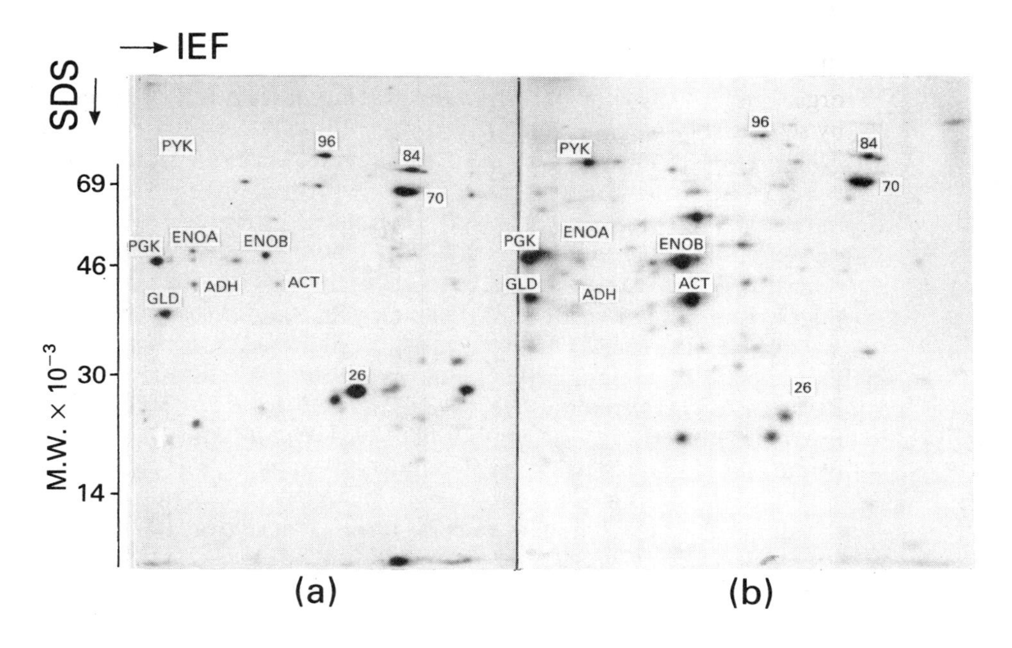

Fig. 3.9 — The synthesis of proteins by yeast cells when subjected to heat shock stress. (a) is a two-dimensional gel of the proteins of *S. cerevisiae* strain MD40-4c (Dobson *et al.* 1982) that become labelled by [^{3}H] leucine 10–30 min after a severe (25°C to 42°C) heat shock. (b) is a similar separation of MD40-4c proteins from non-stressed cells labelled for 20 min at 25°C. Gel fractionation of proteins (by G. Reid and P. W. Piper) was by isoleletric focusing (IEF) in the first dimension and 12.5% polyacrylamide gel in SDS in the second dimension (Piper *et al.* 1986). After heat shock there is a dramatic increase in the synthesis of heat shock proteins HSP26 (spot 26), HSP70 (spot 70) HSP84 (spot 84), and HSP96 (spot 96). There is also considerable synthesis of the glycolytic enzymes phosphoglycerate kinase (PGK), enolase (ENOA and ENOB), and glyceraldehyde phosphate dehydrogenase (GLD), but reduced synthesis of actin (ACT), alcohol dehydrogenase 1 (ADH), and pyruvate kinase (PYK). Identification of enzymes on the gels was from Brousse *et al.* (1985).

there is the immediate cessation of synthesis of many proteins. Simultaneously there is a switch-on of genes for heat shock proteins while certain glycolytic enzymes continue to be made efficiently (Fig. 3.9). This high-temperature synthesis of proteins increases the heat tolerance of the cells possibly because certain heat shock proteins limit damage caused by protein denaturation, and the increased synthesis of glycolytic enzymes may assist in maintenance of energy balance (Piper *et al.* 1986). It reflects both transcriptional changes and a preferential translation of certain mRNAs at high temperature (McGarry & Lindquist 1985). This one stress response therefore involves controls over both promoters and protein synthesis. It will be fascinating to learn whether responses to DNA-damaging agents are equally complex.

REFERENCES

Abraham, J., Nasmyth, K., Strathern, J. N., Klar, A. J. S., & Hicks, J. B. (1984) *J. Mol. Biol.* **176** 307–331.
Ammerer, G., Sprauge, G. F., & Bender, A. (1985) *Proc. Natl. Acad. Sci. USA* **82** 5855—5859.
Bajwa, W., Meyhack, B., Rudolph, H., Schweingruber, A. M., & Hinnen, A. (1984) *Nucl. Acids. Res.* **12** 7721–7739.
Beggs, J. D. (1978) *Nature* **275** 104–109.
Beggs, J. D., Van den Berg, A., van Ooyen, A., & Weissman, C. (1980) *Nature* **283** 835–840.
Beier, D. R. & Young, E. T. (1982) *Nature* **300** 724–728.
Bennetzen, J. L. & Hall, B. D. (1982) *J. Biol. Chem.* **257** 3018–3026.
Borrelli, E., Hen, R., & Chambon, P. (1984) *Nature* **312** 608–612.
Brand, A. H., Breeden, L., Abraham, J., Sternglanz, R., & Nasmyth, K. (1985) *Cell.* **41** 41–48.
Breathnach, R. & Chambon, P. (1981) *Ann. Rev. Biochem.* **50** 349–383.
Brent, R. (1985) *Cell.* **42** 3–4.
Brent, R. & Ptashne, M. (1984) *Nature* **312** 612–615.
Brousse, M., Bataillé, N., & Boucherie, H. (1985) *Appl. Env. Microbiol.* **50** 951–957.
Carlson, M. & Botstein, D. (1982) *Cell.* **28**, 145–154.
Carlson, M., Osmond, B. C., & Botstein, D. (1981) *Genetics* **98** 25–40.
Casadaban, M. & Cohen, S. N. (1980) *J. Mol. Biol.* **138** 179–207.
Clarke, L. & Carbon, J. (1980) *Nature* **257** 504–509.
Cohen, J. D., Goldenthal, M. J., Chow, J., Buchferer, B., & Marmur, J. (1985) *Mol. Gen. Genet.* **200** 1–8.
Darnell, J. E. (1982) *Nature* **297** 365–371.
DiLauro, R., Taniguchi, T., Musso, R., & de Crombrugghe, B. (1979) *Nature* **279** 494–500.
DiMauro, E., Camilloni, G., Della Seta, S., Ficca, A. G., & Negri, R. (1985) *J. Biol. Chem.* **260** 152–159.
Dobson, M. J., Tuite, M. F., Roberts, N. A., Kingsman, A. J., Kingsman, S. M., Perkins, R. E., Conroy, S. C., Dunbar, B., & Fothergill, L. A. (1982) *Nucl. Acids. Res.* **10** 2625–2637.
Domdey, H., Apostol, B., Lin, R. J., Newman, A., Brody, E., & Abelson, J. (1984) *Cell.* **39** 611–621.
Donahue, T. F., Davies, R. S., Lucchini, G., & Fink, G. R. (1983) *Cell.* **32** 89–98.
Entian, K. D. & Frohlich, K. U. (1984) *J. Bact.* **158** 29–35.
Entian, K. D., Hilberg, F., Opitz, H., & Mecke, D. (1985) *Mol. Cell. Biol.* **5** 3035–3040.

Entian, K. D., Kopetzki, E., Frohlich, K. U., & Mecke, D. (1984) *Mol. Gen. Genet.* **198** 50–54.
Faye, G., Leung, D. W., Tatchell, K., Hall, B. D., & Smith, M. (1981) *Proc. Natl. Acad. Sci. USA* **78**, 2258–2262.
Fields, S. & Herskowitz, I. (1985) *Cell* **42** 923–930.
Giniger, E., Varnum, S. M., & Ptashne, M. (1985) *Cell.* **40** 767–774.
Guarente, L. (1984) *Cell.* **36** 799–800.
Guarente, L. & Hoar, E. (1984) *Proc. Natl. Acad. Sci. USA* **81** 7860–7864.
Guarente, L., Lalonde, B., Gifford, P., & Alani, E. (1984) *Cell.* **36** 503–511.
Guarente, L. & Mason, T. (1983) *Cell.* **32** 1279–1286.
Guarente, L. & Ptashne, M. (1981) *Proc. Natl. Acad. Sci. USA* **78** 2199–2203.
Guarente, L., Yocum, R., & Gifford, P. (1982) *Proc. Natl. Acad. Sci. USA* **79** 7410–7414.
Hall, M. N., Hereford, L., & Herskowitz, L. (1984) *Cell.* **36** 1057–1065.
Henikoff, S., Kelly, J. D., & Cohen, E. H. (1983) *Cell.* **33** 607–614.
Hereford, L. M., Osley, M. A., Ludwig, J. R., & McLaughlin, C. S. (1981) *Cell.* **24** 367–375.
Herskowitz, I. & Oshima, Y. (1981) In: *The molecular biology of the yeast Saccharomyces: life cycle and inheritance* (Strathern, J. N., Jones, E. W., & Broach, J. R. eds.) Cold Spring Harbor Laboratory, New York, pp. 181–209.
Hinnebusch, A. G. (1985) *Mol. Cell. Biol.* **5** 2349–2360.
Hinnebusch, A. G. & Fink, G. (1983a) *Proc. Natl. Acad. Sci. USA* **80** 5374–5378.
Hinnebusch, A. G. & Fink, G. (1983b) *J. Biol. Chem.* **258** 5238–5247.
Hinnebusch, A. G., Lucchini, G., & Fink, G. R. (1985) *Proc. Natl. Acad. Sci. USA* **82** 498–502.
Hinnen, A., Hicks, J. B., & Fink, G. R. (1978) *Proc. Natl. Acad. Sci. USA* **75** 1929–1933.
Hong, S. H., Chow, T., Hegde, M. V., & Marmur, J. (1984) *Proc. Natl. Acad. Sci. USA* **81** 2811.
Hope, I. A. & Struhl, K. (1985) *Cell.* **43** 177–188.
Hopper, J., Broach, J., & Rowe, L. (1978) *Proc. Natl. Acad. Sci. USA* **75** 2878–2882.
Iborra, F., Francingues, M. C., & Guerineau, M. (1985) *Mol. Gen. Genet.* **199** 117–122.
Ito, H., Fukuda, Y., Murata, K., & Kimura, A. (1983) *J. Bact.* **153** 163–168.
Jayaram, M., Li, Y–Y, & Broach, J. R. (1983) *Cell.* **34** 95–104.
Johnson, A. D. & Herskowitz, I. (1985) *Cell.* **42** 237–247.
Johnson, W. C., Moran, C. P., & Losick, R. (1983) *Nature* **302** 800–804.
Johnston, S. & Hopper, J. (1982) *Proc. Natl. Acad. Sci. USA* **79** 6971–6975.
Kaback, D. B. & Feldberg, L. R. (1985) *Mol. Cell. Biol.* **4** 751–761.
Kaneko, Y.,
tamai, Y., Toh-e, A., & Oshima, Y. (1985) *Mol. Cell. Biol* **5** 248–252.
Kaufer, N. F., Simanis, V., & Nurse, P. (1985) *Nature* **318** 78–80.

Keller, E. B. & Noon, W. A. (1984) *Proc. Natl. Acad. Sci. USA* **81** 7417–7420.

Keng, T. & Guarente, L. (1986) *Mol. Cell. Biol.* (in press).

Kinnaird, J. H. & Fincham, J. R. S. (1983) *Gene* **26** 253–260.

Kingsman, S. M., Kingsman, A. J., Dobson, M. J., Mellor, J., & Roberts, N. A. (1985) *Biotechnology & Genetics Engineering Reviews* **3** 377–416.

Kingston, R. E., Baldwin, A. S., & Sharpe, P. A. (1985) *Cell.* **41** 3–5.

Lalonde, B., Prezant, T., & Guarente, L. (1986) *Cell.* (in press).

Langford, C. J., Flinz, F-J., Donath, C., & Gallwitz, D. (1984) *Cell.* **36** 645–653.

Langford, C. J. & Gallwitz, D. (1983) *Cell.* **33** 519–527.

Langford, C. J., Nellan, J., Niessing, J., & Gallwitz, D. (1983) *Proc. Natl. Acad. Sci. USA* **80** 1496–1500.

Laughon, A., Driscol, R. M., Wills, N., & Gesteland, R. F. (1984) *Mol. Cell. Biol.* **4** 268–275.

Laughon, A. & Gesteland, R. F. (1982) *Proc. Natl. Acad. Sci. USA* **79** 6827–6831.

Manley, J. L. (1983) *Prog. Nucl. Acid. Res. Mol. Biol.* **30** 195–237.

Matsumoto, K., Adachi, Y., Toh-e, A., & Oshima, Y. (1980) *J. Bact.* **141** 508–527.

Matsumoto, K., Toh-e, A., & Oshima, Y. (1981) *Mol. Cell. Biol.* **1** 83–93.

Meinkoth, J. & Wahl, G. (1984) *Anal. Biochem.* **138** 267–284.

Miller, A. M., MacKay, V. L., & Nasmyth, K. A. (1985) *Nature* **314** 598–603.

Miller, A. M. & Nasmyth, K. A. (1984) *Nature* **312** 247–251.

McAlister, L. & Holland, M. J. (1982) *J. Biol. Chem.* **257** 7181–7188.

McGarry, T. J. & Lindquist, S. (1985) *Cell.* **42** 904–911.

McNeil, J. B. & Smith, M. (1985) *Mol. Cell. Biol.* **5** 3545–3551.

Nasmyth, K. (1985a) *Cell.* **42** 213–223.

Nasmyth, K. (1985b) *Cell.* **42** 225–235.

Nasmyth, K. A., Tatchell, K., Hall, B., Astell, C. R., & Smith, M. (1981) *Nature* **289** 244–250.

Neidhardt, F. C., VanBogelen, R. A., & Vaughn, V. (1984) *Ann. Rev. Genet.* **18** 295–329.

Nogi, Y., Matsumoto, K., Toh-e, A., & Oshima, Y. (1977) *Mol. Gen. Genet.* **152** 137–144.

Oshima, Y. (1982) In: *The molecular biology of the yeast Saccharomyces* (Strathern, J. N., Jones, E. W., & Broach, J. R., eds) Cold Spring Harbor Laboratory, Cold Spring Harbor, New York. pp. 159–180.

Perlman, D. & Hopper, J. E. (1979) *Cell.* **16** 89–95.

Peterson, T. A., Prakash, L., Prakash, S., Osley, M. A., & Reed, S. I. (1985) *Mol. Cell. Biol.* **5** 226–235.

Pikielny, C. W., Teem, J. L., & Rosbash, M. (1983) *Cell* **34** 395–403.

Pinkham, J. L. & Guarente, L. (1985) *Mol. Cell. Biol.* **5** 3410–3416.

Piper, P. W., Curran, B., Davies, M. W., Lockheart, A. & Reid, G. (1986) *Eur. J. Biochem.* **161** 525–531.

Roeder, G. S., Beard, C., Smith, M., & Keranen, S. (1985) *Mol. Cell. Biol.* **5** 1543–1553.

Rolfe, M., Spanos, A., & Banks, G. (1986) *Nature* **319** 339–340.

Rosbach, M., Harris, P. K. W., Woolford, J. L., & Teem, J. L. (1981) *Cell.* **24** 679–687.

Rose, M., Casadaban, M. J., & Botstein, D. (1981) *Proc. Natl. Acad. Sci. USA* **78** 2460–2464.

Rothstein, R. J. (1983) In *Methods in Enzymology* (Wu, R., Grossman, L., & Moldave, K. eds) **101** pp. 202–211, Academic Press, New York.

Ruby, S. W. & Szostak, J. W. (1985) *Mol. Cell. Biol.* **5** 75–84.

Russell, D. W., Smith, M., Cox, D., Williamson, V. M., & Young, E. T. (1983) *Nature* **304** 652–654.

Sancar, G. B., Sancar, A., Little, J. W., & Rupp, D. W. (1982) *Cell.* **28** 523–530.

Santiago, A., Purvis, I. J., Bettany, A. J. E. & Brown, A. A. (1986) *Nucl. Acids Res.,* **14** 8347–8360.

Sarokin, L. & Carlson, M. (1985) *Mol. Cell. Biol* **5** 2521–2526.

Schurr, A. & Yagil, E. (1971) *J. Gen. Microbiol.* **65** 291–298.

Sharpe, P. A. (1985) *Cell* **42** 397–400.

Shine, J. & Dalgarno, L. (1975) *Nature* **254** 34–38.

Siliciano, P. G. & Tatchell, K. (1984) *Cell.* **37** 969–978.

Silver, P. A., Keegan, L. P., & Ptashne, M. (1984) *Proc. Natl. Acad. Sci. USA* **81** 5951–5955.

Smith, M., Leung, D. W., Gillam, S., Astell, C. R., Montgomery, D. L., & Hall, B. D. (1979) *Cell.* **16** 753–761.

Sprague, G. F., Blair, L. C., & Thorner, J. (1983) *Ann. Rev. Microbiol.* **37** 623–660.

Sprague, G. F., Jensen, R., & Herskowitz, I. (1983) *Cell.* **32** 409–415.

Sripati, C. E., Groner, Y., & Warner, J. R. (1976) *J. Biol. Chem.* **251**, 2898–2904.

St John, T. & Davis, R. (1981) *J. Mol. Biol.* **152** 285–315.

Strathern, J., Hicks, J., & Herskowitz, I. (1981) *J. Mol. Biol.* **147** 357–372.

Struhl, K. (1982a) *Nature* **300** 284–287.

Struhl, K. (1982b) *Proc. Natl. Acad. Sci. USA* **79** 7385–7389.

Struhl, K. (1983) *Nature* **305** 391–396.

Struhl, K. (1984) *Proc. Natl. Acad. Sci. USA* **81** 7865–7869.

Struhl, K., Stinchcomb, D. T., Scherer, S., & Davis, R. W. (1979) *Proc. Natl. Acad. Sci. USA* **76** 1035–1039.

Summer-Smith, M., Bozzato, R. P., Skipper, N., Wayne Davis, R., & Hopper, J. E. (1975) *Gene* **36** 333–340.

Tatchell, K., Nasmyth, K. A., Hall, B. D., Astell, C., & Smith, M. (1981) *Cell.* **27** 25–35.

Teem, J., Rodriguez, J. R., Tung, L., & Rosbash, M. (1983) *Mol. Gen. Genet.* **192** 101–107.

Thireos, G., Driscoll Penn, M., & Greer, H. (1984) *Proc. Natl. Acad. Sci. USA.* **81** 5096–5100.

Valenzuela, P., Medina, A., Rutter, W. J., Ammerer, G., & Hall, B. D. (1982) *Nature* **298** 347–350.

Walker, G. C. (1984) *Microbiol. Revs.* **48** 60–93.

Warner, J. R., Gopa, M., Schwindinger, W. F., Studeny, M., & Fried, H. M. (1985) *Mol. Cell. Biol.* **5** 1512–1521.

Watts, F., Castle, C., & Beggs, J. D. (1983) *EMBO J.* **2** 2085–2091.

Wood, C. R., Boss, M. A., Kenton, J. M., Calvert, J. E., Roberts, N. A., & Emtage, J. S. (1985) *Nature* **314** 446–449.

Wright, C. F. & Zitomer, R. S. (1985) *Mol. Cell. Biol.* **5** 2951–2958.

Yamashita, I. & Fukui, S. (1985) *Mol. Cell. Biol.* **5** 3069–3073.

Zalkin, H. & Yanofsky, C. L. (1982) *J. Biol. Chem.* **257** 1491–1500.

Zaret, K. S. & Sherman, F. (1984) *J. Mol. Biol.* **176** 107–135.

4

DNA repair and mutagenesis in *Saccharomyces cerevisiae*

Dr. A. J. Cooper[1] and S. L. Kelly[2]
1. Department of Pathology, Stanford Medical School, Stanford University, California, USA
2. Wolfson Institute of Biotechnology, The University, Sheffield

4.1 INTRODUCTION

The processes of mutagenesis and DNA repair have attracted considerable interest largely because of the need for greater understanding of genotoxic hazards which may increase cancer in the present generation and genetic diseases subsequently. Mutagenesis and DNA repair are best understood in *Escherichia coli* where the modern techniques of molecular biology have provided invaluable insights, and the opportunity exists to extend knowledge of mutagenesis and DNA repair in the eukaryote *Saccharomyces cerevisiae* to the same degree of detail.

The study of mutagenesis and DNA repair in *Saccharomyces cerevisiae* has been led by genetic analysis through the isolation of mutants affected in mutagenesis and DNA repair and by analysing mutation using defined markers for DNA chadces. DNA repair mutants were isolated on the basis of hypersensitivity to the lethal effects ob mutagenic agents. Interesting analogies have already been revealed to the human situation, in contrast to *E. coli,* such as in the number of genes implicated in the DNA repair process of incision at the sites of UV-induced damage (Friedberg 1985a). It as also been found that the gene *RAD*6 plays a central role in induced mutagenesis for almost all mutagens which have been exemined, and that yeast differs considerably from the genetic control elucidated for induced mutagenesis in *E. coli* (Prakash 1976a,b).

In contrast to these mutant studies, which are similar to those applied to *E. coli, Saccharomyces cerevisiae* also offers the capacity to probe aspects of potential importance for genetic change in eukaryotes. In particular, questions may be asked about the effect of chromatin structure; of the differences between nuclear and mitochondrial susceptibility; the variations of sensitivity during the mitotic and meiotic cell cycles and about mechanisms of chromosome loss. The biochemical background to mutagenesis and DNA repair lags bdind the genetic analysis, but this is being remedied particularly through gene cloning studias. A particular advantage of yeast in these studies is the facility for gene disruption and displacement at the chromosomal locus (see Orr-Weaver *et al.* 1983) which will enable detailed and biologically relevant aspects of yeast gene structure, regulation, and function to be analysed.

This chapter is designed to provide a review of the status of research into mutagenesis and DNA repair in *Saccharomyces cerevisiae*, areas of which have been covered elsewhere (Lemontt 1980, Haynes & Kunz 1982, Lawrence 1982, Game 1984).

4.2 GENERAL CONCEPTS OF DNA REPAIR AND MUTAGENESIS

Mutagenesis encompasses all *de novo* heritable changes and can involve gross chromosomal changes as well as changes at the gene level. One disadvantage to the yeast system is the inability to observe individual chromosomes so that aspect of chromosome damage and repair are difficult to study. A promising development in this regard is the technique of orthogonal field electrophoresis which enables fractionation of yeast chromosomes (Schwartz & Cantor 1984).

A conceptual framework for mutagenesis and DNA repair has been derived from research on bacteriophage and *E. coli* which indicates the interrelationship of mutagenesis and DNA repair for some types of DNA damage. DNA repair was proposed by Witkin (1967) to be either accurate (error-free) or inaccurate (error-prone). The latter type could repair DNA damage and restore the integrity of the DNA, but in the process incorporate alterations in the coding sequence to give rise to mutations. Distinct from mutations which require misrepair activity are mutations caused by forms of DNA damage which alter the pairing properties of bases and may give rise to mispairing at DNA replication. This framework may be interpreted as describing DNA repair as accurate or misrepair, while mutagenesis mechanisms may be viewed in terms of misreplication or misrepair.

Before considering aspects of mutagenesis (section 4.11) the processes of DNA repair will be discussed. For a detailed review of DNA repair systems readers are referred to Friedberg (1985a). The mechanisms of DNA repair in *E. coli* are best understood for damage induced by UV radiation and more recently for alkylating agents. The classical DNA lesion studied in UV-induced damage is the pyrimidine dimer, but recently another lesion — the TC(6-4)photoproduct — has been associated with hotspots of mutation in the *lac* I gene of *E. coli* (Brash & Haseltine 1982). The importance of this lesion for mutagenesis is still controversial (Kunz & Glickman 1984), and studies in yeast will be of interest to provide a eucaryote perspective.

The simplest repair pathway for pyrimidine dimers which appears in *E. coli* is photoreactivation where a photolyase catalyses the break of linked pyrimidines in a process dependent on light (Fig. 4.1). Similar activities have

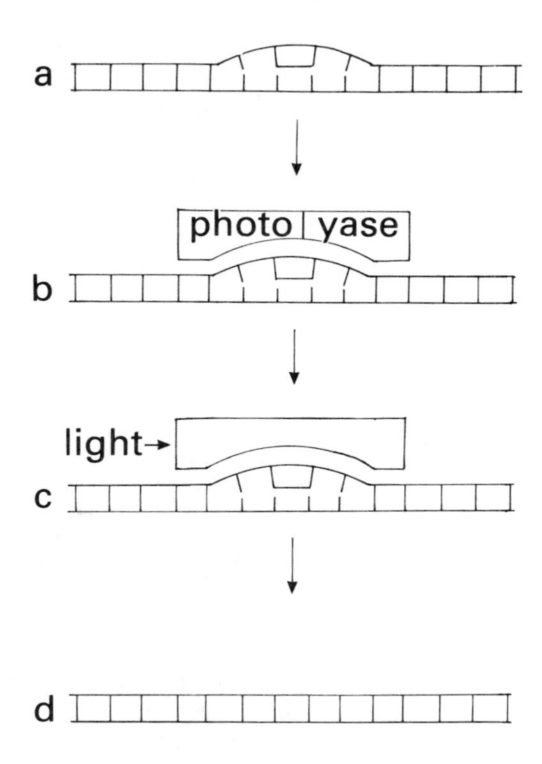

Fig. 4.1 — The photoreactivation of pyrimidine dimers: (a) pyrimidine dimer induced in DNA; (b) interaction of dimer and photoreactivation enzyme; (c) absorption of light of wavelength less than 300 nm; (d) release of enzyme and restoration of the DNA.

been detected in other systems up to, and including, placental mammals, although the occurrence of photoreactivation in these systems remains controversial. Excision repair is a major repair pathway for bulky DNA lesions such as pyrimidine dimers as well as for other lesions in the DNA such as apurinic or apyridimic (AP) sites. Excision processes involve the recognition of the damage, incision into the phosphodiester backbone, removal of the lesion in a single strand of the DNA, and the resynthesis and religation of a corrected strand using the complementary strand as a template (Fig. 4.2).

Processes of DNA metabolism also act on DNA which contains lesions and is being actively synthesized. These processes are often described as being DNA repair systems although they represent mechanisms of damage tolerance during replication and do not result in the removal of the damage.

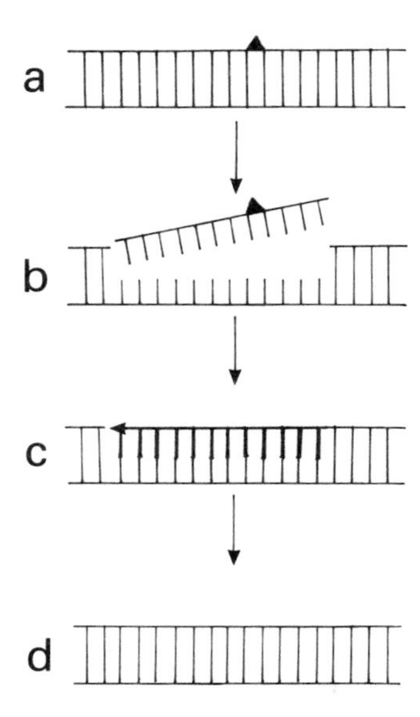

Fig. 4.2 — Short patch excision repair of pyrimidine dimer damage in *E. coli*: (a) pyrimidine dimer induced in DNA; (b) incision by endonuclease produces cleavage on either side of dimer; (c) dimer removed and new strand resynthesized by DNA polymerase using the opposite strand as a template; (d) religation to restore intact double helix. A long patch excision mechanism also operates which involves expansion of the excision gap prior to resynthesis.

The presence of bulky DNA lesions such as pyrimidine dimers results in the inhibition of DNA synthesis and is best understood again in *E. coli* systems. Single strand gaps are left in the daughter strand of the DNA opposite the lesion, and these are filled in a process involving transfer of DNA from the sister duplex which has been completed (Fig. 4.3).

Study of alkylating damage in *E. coli* has revealed additional repair mechanisms for the removal of alkylated bases and for the removal of alkyl adducts on bases. In particular, systems exist for removing bases (glycosylases) from their position in the DNA to leave apurinic or apyridimic (AP) sites which may be recognised by an AP endonuclease. This introduces a nick into the DNA backbone, and such lesions may then be subject to excision repair. The other system for the repair of alkylation repair involves the inducible *E. coli* repair process known as the adaptation response. In this an alkyl transferase removes methyl, or less efficiently ethyl, groups from 0^6 alkyl-guanine and avoids the mispairing potential of the altered base at DNA replication. The methyl or ethyl group is transferred from the guanine base onto a cysteine residue of the transferase.

The other DNA repair pathway which has been elucidated, but which is relatively poorly understood at the biochemical level, is that of mismatched base repair. This repair pathway in *E. coli* is able to distinguish parental from daughter strands of the DNA following DNA replication because of transient undermethylation of the daughter strand and to correct mis-

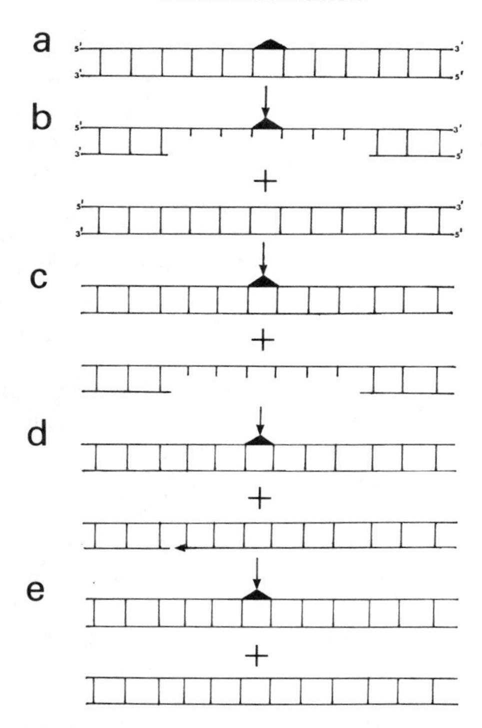

Fig. 4.3 — Post-replication repair of pyrimidine dimer damage; (a) pyrimidine dimer induced in DNA; (b) DNA replication bypasses the damaged strand, but leaves single strand gap opposite the pyrimidine dimer; (c) Non-reciprocal recombinational event from the sister-duplex fills the single-strand gap; (d) the gap left by the non-reciprocal transfer is filled, using the opposite strand as a template; (e) replication past the pyrimidine dimer is completed.

matched bases introduced into the daughter strand. Mismatched base repair is also important in recombination where heterozygous regions occur, although the eukaryote repair mechanisms are unknown.

4.3 DNA REPAIR IN *SACCHAROMYCES CEREVISIAE*

The repair mechanisms outlined above are founded in the *E. coli* studies, but some similar activities have been found in eukaryote systems. The processes of DNA repair in *Saccharomyces cerevisiae* represent the most convenient eukaryote system to investigate because of the ease of manipulation, the classical genetic system for the isolation and analysis of mutants, and the ability to clone yeast genes and to return wild-type and *in vitro* mutagenized genes into yeast to probe aspects of regulation, structure, and activity.

4.4 PHOTOREACTIVATION

The simplest repair mechanism known is the enzymatic photoreactivation of UV-induced pyrimidine dimers. A process by which an enzyme, stimulated by light, can directly reverse the formation of pyrimidine dimers. This phenomenon is found in a wide variety of organisms (Friedberg 1985a), including *S. cerevisiae* (Resnick 1969, Resnick & Setlow 1972). There are at least two genes in *S.cerevisiae* which are involved in photoreactivation, *PHR*1 and *PHR*2 (Resnick 1969, MacQuillan *et al.* 1981), and they appear to be loosely linked (MacQuillan *et al.* 1981). This is probably the reflection of at least two separate photoreactivating enzymes in yeast since the various attempts to isolate the enzyme activity from wild-type cells has led to the identification of different proteins. Muhammed (1966) isolated a 30 kDa protein together with a low molecule weight component; neither component absorbed light at the optimal wavelength for photoreaction: 350 nm (Setlow & Boling 1963). Boatwright *et al.* (1975) have isolated a 130 kDa protein which could be resolved into 2 proteins of 60 kDa and 85 kDa, both of which were necessary for the photolyase activity. This enzyme also showed no ability to absorb light in the photoreactivating range, although it was found to be associated with a fluorescent 450 dalton molecule, termed activator III, which may be the chromophore (Werbin & Madden 1975, 1977). Iwatsuki *et al.* (1980) have isolated a 51 kDa protein which absorbs light at 380 nm and is associated with a low molecular weight flavin moiety. In addition to these findings Madden & Werbin (cited in Schild *et al.* 1984) report the existence of 2 separate photoreactivating enzymes: a monomer and a heterodimer, both of which require a cofactor.

The *PHR*1 gene has been cloned and sequenced (Yasui & Chevallier 1983, Schild *et al.* 1984, Yasui *et al.* 1984), and it shows extensive homology with the *E. coli phr* gene (Yasui *et al.* 1984). Langeveld *et al.* (1985) have also transformed photoreactivation deficient yeast with a plasmid carrying the *E. coli phr* gene and observed complementation of the yeast defect.

The other repair processes in yeast are termed dark repair processes since unlike photoreactivation they are not light-dependent.

4.5 RADIATION-SENSITIVE (*RAD*) MUTANTS: ASSIGNMENT TO DNA REPAIR PATHWAYS

Most of the *S.cerevisiae* DNA repair mutants were originally isolated by their sensitivity to radiation (Nakai & Matsumato 1967, Snow 1967, Cox & Parry 1968, Resnick 1969, Brown & Kilbey 1970, Game & Cox 1971). These radiation-sensitive (*rad*) mutants have been arranged into 3 epistasis groups for the dark repair of radiation damage (Kahn *et al.* 1970, Game & Cox 1972, 1973). These epistasis groups are based on whether mutations in two *RAD* genes exert an epistatic, synergistic, or additive effect on radiation sensitivity when they are present in the same strain. If the two mutant genes act in the same repair pathway, then the double mutant would be no more or less sensitive than the most sensitive single mutant, i.e. the interaction would be epistatic. If the two mutant genes represent functions required for different repair pathways then the sensitivity of the double mutant would be equal to the additive sensitivities of the single mutants, i.e. an additive interaction. A synergistic inter-reaction occurs when the two repair pathways blocked in the double mutant can act on the same damage. Thus, in either of the single mutants, the intact pathway can partially compensate for the blocked pathway, and the double mutant would consequently be more sensitive than in an additive interaction. The three epistasis groups for radiation-induced damage are designated the *RAD*3, *RAD*6, and *RAD*52 groups after dominant loci in the groups (Fig. 4.4).

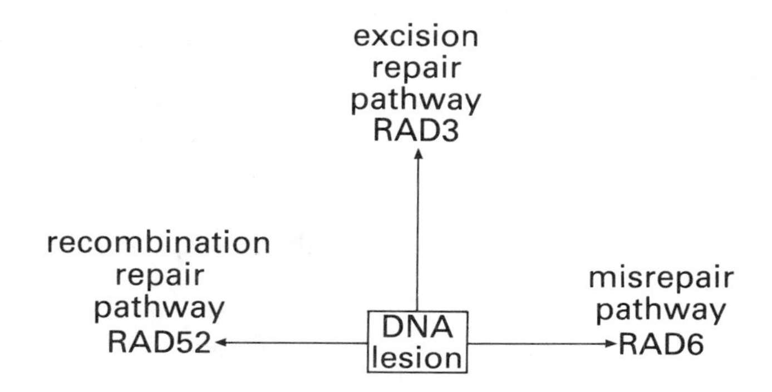

Fig. 4.4 — Major DNA repair pathways in *Saccharomyces cerevisiae*.

4.6 THE *RAD*3 EPISTASIS GROUP INVOLVED IN EXCISION REPAIR

4.6.1 Known phenotypic effects of the mutants

The *RAD*3 group currently consists of the *RAD*1, *RAD*2, *RAD*3, *RAD*4, *RAD*7, *RAD*10, *RAD*14, *RAD*16, *RAD*23, *RAD*24, *MMS*19, *CDC*8, and *CDC*9 genes. This epistasis group is responsible for the excision repair of pyrimidine dimers (Unrau *et al.* 1971, Resnick & Setlow 1972, Waters & Moustacchi 1974a, Prakash 1975a,b, 1977a,b, Reynolds 1978, Prakash & Prakash 1979, Prakash *et al.* 1979, Reynolds & Friedberg 1981, Reynolds *et al.* 1981, Wilcox & Prakash 1981, Miller *et al.* 1982a,b). The *CDC*8 and *CDC*9 genes have been cloned and the gene products identified: *CDC*9 codes for DNA ligase (Johnston & Nasmyth 1978, Barker & Johnston 1983, Barker *et al.* 1984), and the *CDC*8 gene encodes thymidylate kinase (Kuo & Cambell 1983, Sclafani & Fangman 1984, Birkenmeyer *et al.* 1984). These genes are cell cycle genes which play major roles in DNA metabolism, and this raises an interesting point: that genes involved in repair processes may also be required in other aspects of DNA or cellular metabolism. In addition, of course, some genes may be involved in more than one repair process, for example *CDC*9 also falls into the *RAD*52 group and *CDC*8 is epistatic to *RAD*6 (see later). It is clear to see how an excision repair process and DNA replication might require a DNA ligase activity. An examination of the transcriptional regulation of *CDC*9 shows that *CDC*9 mRNA increases 4- to 5-fold in late G1 phase, after completion of START in the mitotic cell cycle, but before S-phase (Paterson *et al.* 1985). Exposure of cells to UV produces an 8-fold increase in mRNA levels, and a similar increase was seen with methylmethanesulphonate (MMS) (Peterson *et al.*, 1985). It has been suggested that the *CDC*9 induction signal during DNA repair is the presence of single-strand breaks or single-stranded DNA; however, this is unlikely to be the signal in G1 since induction occurs before the commencement of DNA synthesis (Peterson *et al.* 1985).

Yeast mutants defective in thymidylate kinase (*cdc*8 mutants), are defective in chromosomal, mitochondrial, and 2μ plasmid replication at the restrictive temperature. Most alleles also demonstrate lowered UV-induced mutation levels at the permissive temperature when replication is not impaired. It is not clear whether this is due to increased error-free repair such as that controlled by the *RAD*3 group or decreased error-prone repair. In this respect it should be noted that the *CDC*8 gene is also a member of the *RAD*6 epistasis group (Prakash *et al.* 1979) which controls probably all error-prone, as well as some error-free, repair in yeast (see below). It is possible that at the permissive temperature, the thymidylate kinase encoded by some *cdc*8 alleles is still partially defective, so that while replication can proceed, more subtle alterations in nucleotide pools affect the quality of repair replication. In accordance with this idea it should be noted that a yeast mutant deficient in deoxycytidylate (dCMP) deaminase shows lower spontaneous mutation rates at some loci and higher rates for other loci (Maus & Haynes 1984). The *cdc*8 mutation was found to decrease UV-induced reversion at four alleles: *lys*2-1, *arg*4-17, *tyr*1 and *ura*1 (Prakash *et al.* 1979).

Specific pyrimidine dimer-DNA glycosylases have been used to follow the removal of dimers in *rad*1, *rad*2, *rad*3, *rad*4, *rad*7, *rad*10, *rad*14, *rad*16, *rad*23, and *mms*19 cells (Prakash, 1977a,b, Prakash & Prakash 1979, Reynolds & Friedberg 1981a, Miller *et al.* 1982). The results showed *rad*1, *rad*2, *rad*3, *rad*4, and *rad*10 mutants to be completely defective in pyrimidine dimer removal, while there was some residual repair in *rad*7, *rad*14, *rad*16, *rad*23, and *mms*19 mutants. In addition, *rad*1, *rad*2, *rad*3, and *rad*4 mutants showed no detectable incubation-dependent DNA strand breaks after UV as measured by sedimentation in alkaline sucrose gradients (Reynolds & Friedberg 1980, 1981). This suggests a defect in these mutants in the incision or pre-incision events of excision repair. Confirmation of this comes from the studies of Wilcox & Prakash (1981), who put the *rad* mutations in a DNA ligase deficient *cdc*9 background and measured the accumulation of DNA single-strand breaks at the restrictive temperature. There was no accumulation of breaks in the *rad*1, *rad*2, *rad*3, *rad*4, and *rad*10 mutants, indicating that no incision was taking place, while DNA from the *rad*14 *cdc*9 and *rad*16 *cdc*9 was of lower molecular weight, suggesting that the *RAD*14 and *RAD*16 genes are in a step subsequent to incision. Subsequent to incision, gaps in the DNA would occur because of the DNA ligase deficiencies.

In view of the slight UV-sensitivity of *rad*7, *rad*14, *rad*16, *rad*23, and *mms*19 mutants, together with their residual repair capacities, it may be pertinent to question the role of these genes in the repair process. It is possible that the failure to observe more defective responses in these mutants is a consequence of the available alleles being leaky. Alternatively it is quite possible that these genes encode proteins which enhance the efficiency of the repair process without being absolutely required. An examination of the phenotypes of relevant deletion mutants should shed some light on this situation. The *rad*24 mutant, although included in the *RAD*3 group, has not been examined biochemically for excision repair capacity.

Wild-type yeasts do not repair pyrimidine dimers from the mitochondria (Waters & Moustacchi 1974b, Prakash 1975), but repair of UV-irradiated plasmid DNA does occur. A comparison of the efficiency of this process in *RAD*+, *rad*1, *rad*2, *rad*3, *rad*4, and *rad*10 strains gave some surprising results (Dominski & Jachymczyk 1984a,b). Only *rad*1 mutants were defective in repair of plasmid DNA as measured by the ability of the plasmid to transform. This was presumed to be a consequence of the involvement of the *RAD*2, *RAD*3, *RAD*4, and *RAD*10 genes in the recognition of dimers in chromatin, or in making such damage accessible to repair enzymes. This was an attractive explanation because of the analogy with human cell studies which had previously shown that under conditions in which normal human fibroblast extracts could remove dimers from chromatin and naked DNA, cell extracts from *Xeroderma pigmentosum* groups A, C, and G removed dimers from naked DNA only while XP-D cells were defective in dimer removal from both substrates (Mortelmans *et al.* 1976, Friedberg *et al.* 1979, Fujiwara & Kano 1983). The human disease *Xeroderma pigmentosum* involves a defect in excision repair (Arlett & Lehmann 1978, Friedberg 1985a,b). However, in another study (White & Sedgwick 1985), *rad*1-1, *rad*2, *rad*3-2, *rad*4-4, *rad*7-1, and *rad*14 mutants were all found to be defective in the repair of UV-irradiated plasmid DNA as measured by transformation efficiency.

4.6.2 Recombinant DNA studies of the *RAD*1epistasis group genes

The *RAD*1, *RAD*2, *RAD*3, *RAD*4, *RAD*7, and *RAD*10 genes have all been cloned at the time of writing, and the *RAD*1, *RAD*2, *RAD*3, *RAD*7, and *RAD*10 genes have been sequenced (Naumovski & Friedberg 1982, 1983a,b, Higgins *et al.* 1983a,b, Higgins *et al.* 1984, Yang & Friedberg 1984, Naumovski *et al.* 1985, Nicolet *et al.* 1985, Weiss & Friedberg 1985, Reynolds *et al.* 1985a,b, R. Fleer *et al.* 1986 — manuscript in preparation, Perozzi & Prakash 1986). The *RAD*1, *RAD*2, *RAD*3, and *RAD*10 genes were all cloned through complementation of UV sensitivity by multicopy plasmids carrying yeast genomic DNA. The *RAD*4 gene could not be isolated in this way, and it eventually became evident that plasmids carrying a functional *RAD*4 gene could not be propagated in *E. coli* (R. Fleer, C. Nicolet, G. Pure, & E. C. Friedberg — in preparation). Whether this is due to the presence of a specific DNA sequence in the *RAD*4 gene, or to the possible lethal effect of *RAD*4 protein in *E. coli*, is as yet unknown. The *RAD*4 gene has consequently been cloned by gap repair of specifically gapped *RAD*4 containing plasmids (R. Fleer, C. Nicolet, G. Pure, & E. C. Friedberg — in preparation).

As already mentioned, *rad*7 mutants are not as UV sensitive as other mutants in the *RAD*3 group, and cloning this gene by the complementation of UV sensitivity would be difficult because of the small window of sensitivity to work within. This problem has been overcome for the *RAD*7 gene which has been cloned by chromosome walking from the nearby *CYC*1 and *OSM*1 genes (Perozzi & Prakash 1986). These three genes comprise the COR gene cluster on chromosome X, and interestingly they may be related to the genes *ANP*1-*RAD*23-*CYC*7 (McKight *et al.* 1981). Accordingly, the *RAD*23 gene is also in the *RAD*3 group and is involved in excision repair. More interestingly, *rad*7 *rad*23 mutants are more UV sensitive and more defective in pyrimidine dimer excision repair than either single mutant, suggesting that the two genes may perform similar functions in repair so that when one gene is defective the other can partially compensate for it (Miller *et al.* 1982). In fact, it has now been shown that a *RAD*7 gene with the first 99 codons deleted complements a strain with the *RAD*7 gene deleted, but not a strain with both the *RAD*7 and *RAD*23 genes deleted (Perozzi & Prakash 1986). The *RAD*7 and *RAD*23 genes do not complement the *rad*23 and *rad*7 deletion mutants respectively when they are on multicopy vectors. Thus, it seems that while the two proteins may share domains for common functions, they also possess some functions that are distinct. The proposal that these two gene clusters are related and have probably arisen by duplication and transposition is very convincing, and it would be interesting to know what the distinct functions of the *RAD*7 and *RAD*23 proteins are, and how they may be evolutionarily related.

The *RAD*1, *RAD*2, *RAD*3, *RAD*7, and *RAD*10 genes have coding regions of 2916 bp, 2925 bp, 2334 bp, 1695 bp, and 588 bp respectively. Like most *Saccharomyces cerevisiae* genes transcribed by RNA polymers II, they apparently contain no introns. The calculated sizes of the proteins that could be encoded by *RAD*1, *RAD*2, *RAD*3, *RAD*7, and *RAD*10 are 110 kd, 111 kd, 89.9 kd, 63.7 kg, and 22.6 kd respectively. The *RAD*4 gene has not been sequenced to date, but insertional mutagenesis studies have located the gene within a 2.3 kbp fragment, suggesting that the *RAD*4 protein could be about 90 kd (R. Fleer, C. Nicolet, G. Pure, & E. C. Friedberg — in preparation).

There are two conserved sequences in the 5′ regions of the *RAD*1, *RAD*2, and *RAD*3 genes (CS1 and CS2) and also a region of homology in the 3′ regions of these genes (Nicolet *et al.* 1985). These sequences are not found in the *RAD*10 gene (Weiss & Friedberg 1985). While there is some evidence for both CS1 and CS2 in some other yeast genes, CS3 has not been detected in any other yeast genes examined (Nicolet *et al.* 1985). *RAD*1 and *RAD*2 also contain regions of dyad symmetry which could form hairpin loops (G*n*-13 kcal/mol), and if such loops formed they would result in the looping out of CS1 in both genes (Nicolet *et al.* 1985).

At the amino acid sequence level there are three brief regions of homology between the predicted *RAD*1 and *RAD*2 proteins. Curiously, these regions occur in reverse order in the two polypeptides although the relative spacing of these regions from each other and from the ends of polypeptides is similar. While these regions are unlikely to have evolved from a common polypeptide they may reflect convergent evolution of polypeptides involved in similar functions (Nicolet *et al.* 1985).

A range of prokaryotic and eukaryotic proteins including some *E. coli* proteins involved in excision repair: *UVR*A and *UVR*C (necessary for incision), *UVR*D (DNA helicase), and *Pol*A (DNA polymerase I) share a region of homology which includes the sequence: Gly-Lys-Thr (Furich & Emmerson 1984, Arikan *et al.* 1986, Backendorf *et al.* 1986, Friedberg *et al.* 1986, Husain *et al.* 1986). This tripeptide is characteristic of proteins which bind guanine/adenine nucleotides and have intrinsic GTPase or ATPase activity and is also present in the predicted amino acid sequences of *RAD*1, *RAD*3, and *RAD*10, suggesting that these proteins may have similar properties. In addition the *RAD*3 and *RAD*10 protein sequences contain homeo boxes suggestive of an α helix-turn-α helix secondary structure characteristic of a DNA binding domain (Naumovski & Friedberg 1986). No extensive homology has been found between any of the sequenced *RAD* proteins and the *UVR* proteins, and expression of the *UVR*A or *UVR*B proteins in *rad*1, *rad*2, *rad*3, *rad*4, or *rad*10 mutants did not restore UV-resistance (Planque *et al.* 1984). Similarly, expression of *RAD*3 or *RAD*10 in *E. coli* does not complement UV sensitivity of *UVR*A, *UVR*B, *UVR*C, or *UVR*D mutants (L. Naumovski, W. Weiss, & E. C. Friedberg — see Friedberg *et al.* 1986). These results may reflect the greater complexity of DNA repair in eukaryotic chromatin. An exciting development comes from the finding that the carboxyl terminal half of the *RAD*10 protein sequence is approximately 30% homologous with the middle third of the human excision repair gene *ERCC*1 protein sequence. Furthermore, within this homologous region there is a stretch of 25 amino acids which shows a 56% level of homology (van Duin *et al.* 1986). The human *ERCC*1 gene was isolated from human genomic DNA which complemented a UV-sensitive Chinese Hamster Ovary (CHO) cell mutant from complementation group 2 (Westerveld *et al.* 1984). There are to date, five CHO complementation groups for UV-sensitivity, and all five are defective in the incision step of excision repair (Thompson *et al.* 1982). Similar transfection experiments to complement the CHO mutants with human DNA have also been reported (Rubin *et al.* 1983, MacInnes *et al.* 1984, Thompson *et al.* 1985a), and it is a real possibility that this approach may lead to the isolation of the human DNA repair genes required for the incision step.

It is a curious observation that as yet no overlap has been found between the CHO complementation groups and the nine *Xeroderma pigmentosum* (XP) complementation groups (Arlett & Lehman 1978, Mashell *et al.* 1983, Fischere *et al.* 1985, Thompson *et al.* 1985b). The disease *Xeroderma pigmentosum* is thought to directly reflect defects in DNA repair genes, and if there is significant homology between repair processes in human and CHO cells, as would be indicated by the *ERCC*1 gene, then one might expect some correlation between these groups of genes. The possibility that the 9

XP genes and the 5 CHO genes reflect the involvement of at least 14 genes in mammalian excision repair seems unlikely in the absence of some bias for the selection of different genes in the two systems. However, as pointed out by Friedberg *et al.* (1986), there may be just such a bias: the XP groups were isolated from XP patients, and it is possible that the failure to isolate defective complementation groups which correspond to the CHO groups is a reflection of the incompatibility of such defects with embryogenesis. The UV-sensitive CHO mutants were isolated by *in vitro* mutagenesis screening protocols which may allow selection of mutants with defects in these genes. There is also evidence that some mutations are preferentially selected during *in vitro* screenings of mutagenized CHO cells, and this is probably due to regions of the CHO genome being in a hemizygous state (Siminovitch 1976). The involvement of at least 14 genes in mammalian excision repair would be in agreement with the 13 genes which have been placed in the *RAD*3 epistasis group for yeast excision repair of UV-induced damage.

The alternative possibility, of course, is that the disease *Xeroderma pigmentosum*, may result in an effect on excision repair capacity, but not represent mutational changes in the genes controlling this repair process.

The *RAD*3 gene is essential for haploid yeast cell viability even in the absence of DNA damage (Naumovski & Friedberg 1983, Higgins *et al.* 1983b). Diploid cells containing one intact and one disrupted copy of the *RAD*3 gene are viable, but following sporulation two viable and two inviable spores are segregated (Naumovski & Friedberg 1983). The essential role of the *RAD*3 gene is presumably distinct from its role in excision repair since *rad*3 mutants have been isolated which are perfectly viable, but which show no residual DNA repair capacity. In addition, an extensive mutagenesis study has demonstrated that the *RAD*3 gene is highly susceptible to inactivation for the repair function by single point mutations, while the essential function is resistant to this kind of inactivation (Naumovski & Friedberg 1986). Mutations at a number of sites including sites within the putative DNA and nucleotide binding domains can result in loss of repair capability. Two conditional lethal mutants for the essential function have been isolated: one which contains a point mutation at codon 73, and one which contains adjacent point mutations in codons 595 and 596 which are within the putative DNA binding box (Naumovski & Friedberg 1986). The differential susceptibility of the two *RAD*3 functions to mutagenic inactivation may be due to different protein–protein or protein–DNA interaction requirements in the two processes. The involvement of *RAD*3 protein in what is most probably a multiprotein complex during excision repair, may mean that mutational changes in many regions of the protein leads to inactivation of the complex. On the other hand the essential function may be more resistant to inactivation because it requires fewer interaction sites in the *RAD*3 protein (Naumovski & Friedberg 1986).

It is intriguing that a DNA repair protein may have a completely distinct function in some essential aspect of DNA or cellular metabolism. It has been suggested that the involvement of *RAD*3 protein in excision repair may limit the availability of the protein for its essential role, following UV-irradiation, thus cell division may be halted, providing opportunity for the removal of DNA damage before replication (Naumovski & Friedberg 1986).

Measurement of the *RAD*1, *RAD*2, *RAD*3, and *RAD*10 mRNA levels have indicated that all four genes are weakly expressed (Naumovski & Friedberg 1982, 1984). This is in accordance with the finding that there is no apparent codon bias in these *RAD* genes (Nicolet *et al.* 1985), a characteristic of other weakly expressed yeast genes (Bennetzen & Hall 1982). The *UVR*A, *UVR*B, and *UVR*D genes are also weakly expressed, but are induced in response to DNA-damaging agents (Casadaban *et al.* 1980, Kenyon & Walker 1980, 1981, Kacinski *et al.* 1981, Foglivano & Schendel 1981, Backendorf *et al.* 1982). The inducibility of the *URV*C gene is still controversial (Backendorf *et al.* 1983, Forster & Strike 1985).

4.7 GENES WHICH ARE INDUCED BY DNA DAMAGE

In parallel to the experiments of Kenyon & Walker (1981), in which Mud (Ap *lac*) operon fusion vectors were used to identify damage-inducible genes in *E. coli,* integrant plasmids carrying gene fusions between the 5′ end of *RAD* genes and the *E. coli lac*Z gene were used to study *RAD* gene regulation (Naumovski & Friedberg 1984, Naumovski *et al.* 1985, Robinson *et al.* 1986). These fusion plasmids express fusion proteins the levels of which can be measured by assaying β-galactosidase activity. Results confirmed that basal *RAD*1-*lac*Z, *RAD*2-*lac*Z, *RAD*3-*lac*Z, and *RAD*10-*lac*Z are very low relative to proteins encoded by strongly expressed yeast genes similarly fused to *lac*Z (Friedberg *et al.* 1986). Levels of the *RAD*1-*lac*Z and *RAD*3-*lac*Z fusion proteins did not increase after exposure to DNA-damaging agents (Naumovski *et al.* 1985, Nagpal *et al.* 1985, Robinson *et al.* 1986). The *RAD*2-*lac*Z fusion, however, was found to be inducible by a variety of DNA-damaging agents: UV, X-rays, 4-nitroquinoline-1-oxide (4NQO), and nalidixic acid, although the increase was only 4–6 fold. Induction of *RAD*2-*lac*Z was confirmed by measuring *RAD*2 mRNA levels (Robinson *et al.* 1986). Lower levels of induction (about 2–3 fold) were observed with bleomycin, nitrogen mustard, mitomycin C, and methylmethanesulphonate (MMS), and no induction was seen with heat shock or amino acid starvation (Robinson *et al.* 1986). The range of agents that can induce the *RAD*2 gene suggests that the induction signal is probably a general one such as an intermediate in the repair process: the presence of single-strand breaks, for example, or possibly a feedback mechanism which responds to a reduction in available repair protein. The finding that *RAD*2 is also inducible in other *RAD* mutants in the *RAD*3 group, however, suggests that it is not the presence of single-strand breaks which is the inducing signal (G.W. Robinson & E. C. Friedberg — personal communication).

The same gene fusion approach has been used to identify other damage-inducible (*DIN*) genes (Ruby & Szostak 1985). Six *DIN* genes have been identified which are induced by UV-irradiation; five of these: *DIN*1, *DIN*2, *DIN*3, *DIN*4, and *DIN*6 are also induced by X-rays, methotrexate, MMS, and *N*-methyl-*N*′-nitro-*N*-nitrosoguanidine (MNNG), while ethylmethane-sulphonate (EMS) induces only *DIN*1, *DIN*3, *DIN*4, and *DIN*6. Thus there are at least three regulatory classes of *DIN* genes. McClanahan & McEntee (1984) have used a different approach to isolate damage-inducible genes from yeast. Labelled cDNA was made from mRNA isolated from control cells or cells exposed to DNA-damaging agents and this was used to screen a yeast library. This technique enabled the identification of damage-inducible genes, i.e. clones which gave a stronger hybridization signal with cDNA from damaged cells, and damage-repressible genes, i.e. clones which gave a stronger hybridization signal with cDNA from control cells. The genes identified in this procedure have been termed *DDR* genes (for DNA damage responsive). Four *DDR* genes have been characterized; two are single copy genes coding for a 3.2 kbp and a 0.5 kbp transcript respectively. The 0.5 kbp transcript increases more than 15-fold within 20 minutes after exposure of cells to 4NQO (McClanahan & McEntee 1984). There is no apparent overlap between the *DIN* and *DDR* genes so far characterized, or between *DIN*1, *DIN*3, *DIN*4, *DIN*6, and the *RAD* genes, *RAD*1, *RAD*2, *RAD*3, *RAD*10, *RAD*6,*RAD*50, *RAD*51, *RAD*52, *RAD*54, and *RAD*55 (Ruby & Szostak 1985). Of course at present it is not known how many, if any at all, of these *DIN* and *DDR* genes are DNA repair genes. The evidence so far implies a complex network of responses following exposure to a variety of damaging agents, which probably results in a complete change in the physiological state of the cell.

It is hoped that the eventual over-expression and isolation of the *RAD* proteins will enable reconstitution of an excision repair complex *in vitro*. The *RAD*10 protein is the only one of these proteins which can be successfully expressed in *E. coli* maxicells (Weiss & Friedberg 1985). Failure to express *RAD*1, *RAD*2, and *RAD*3 in this way may be a consequence of the related 5′ and 3′ sequences in these genes which are not present in the *RAD*10 gene sequence (Nicolet *et al.* 1985). In addition, transcription and/or translation of *RAD*10 may be more efficient in *E. coli* owing to the presence of a Shine–Delgarno sequence (Shine & Delgarno 1974) in which 10 of 13 nucleotides are complementary to the 3′ terminus of the *E. coli* 16S ribosomal RNA (Weiss & Friedberg 1985).

4.8 THE *RAD*6 PATHWAY INVOLVED IN MUTAGENIC REPAIR

All UV-induced mutagenesis in yeast is dependent on a functional *RAD*6 gene, and other mutants within the *RAD*6 group also exhibit reduced levels of UV-induced mutation (Lawrence *et al.* 1970, 1974, Eckardt *et al.* 1975). Mutations in genes from the *RAD*3 group result in some increase in UV-induced mutations in nuclear, but not mitochondrial, DNA (Moustacchi 1969, 1972, Eckardt *et al.* 1975, Eckardt & Haynes 1977), and this is probably a reflection of pyrimidine dimers in nuclear DNA becoming substrates for error-prone repair in the absence of error-free excision repair. The *RAD*6 group currently consists of *RAD*6, *RAD*8, *RAD*9, *RAD*15, *RAD*18, *REV*1, *REV*2, *REV*3, *UMR*1, *UMR*2, *UMR*3, *MMS*3, *CDC*8, *PSO*1, *CDC*7 (Lawrence & Christensen 1976, Lemontt 1977, Prakash *et al.*

1979, McKee & Lawrence 1980, Cassier *et al.* 1981, Njagi & Kilbey 1982), and mutations in any of these genes confer some degree of UV and X-ray sensitivity. In addition, double mutant studies have implied a role for the *RAD*14 gene in this group (Prakash & Prakash 1979). Mutants in the *RAD*6 group show a number of phenotypic differences in response to DNA damage, and although they may all be epistatic to *RAD*6, this group seems to include a number of different repair processes, the biochemical natures of which are unknown. It is clear, however, that post-replication repair following UV-irradation is *RAD*6-dependant (Prakash 1981). Other genes in this group may be involved to differing extents since postreplication repair is only partly inhibited in *RAD*18 and quite normal in *REV*3 (Prakash 1981). Of course, interpretation difficulties arise in the absence of information regarding the potential leakiness of some alleles, but it seems unlikely that this could account for all the phenotypic differences observed in this group. It should be noted that the double mutant studies on which the epistasis groups are based, generally involve the determination of the interaction with only one gene from each group, and it does not follow that a gene which is epistatic to one gene in a given group will also be epistatic to the other genes in that group.

Mutagenesis levels induced by X-rays are also *RAD*6-dependent and, similarly, are defective in most of the other mutants in this group. X-ray mutagenesis is, however, independent of *RAD*9, while both *rad*6 and *rad*9 are X-ray sensitive and *rad*6 *rad*9 double mutants give an epistatic response to X-rays. These results have been interpreted to mean that *RAD*6 and *RAD*9 function in an error-free repair process for X-ray induced damage, while *RAD*6, but not *RAD*9, also functions in an error-prone process for this damage (McKee & Lawrence 1979a,b, Lawrence 1982). Similarly, *rad*18-2 mutants are as sensitive to X-rays as *rad*6 mutants, but show little or no increase in X-ray or UV-induced mutagenesis, suggesting that *RAD*18 functions in only error-free repair (Lawrence 1982).

More convincing evidence that the *RAD*6 gene controls both error-free and error-prone processes comes from the study of suppressors of *rad*6 mutants. The dominant *SRS*2 mutation suppresses the UV-sensitivity of *rad*6-1, *rad*6-3, *rad*18-3, and *rad*18-4 mutants, but does not suppress known amber or ochre suppressible genes for other functions (Lawrence & Christensen 1979). In addition, this suppressor does not suppress the UV-immutability of *rad*6 mutants or the sporulation defect in *rad*6-1 homozygous diploids. Suppression of UV-sensitivity was not found to be a consequence of enhanced excision repair capacity, and it was consequently thought to reflect the existence of *RAD*6 and *RAD*18-dependent error-free repair pathway and a *RAD*6-dependent, but *RAD*18-independent, error-prone repair pathway (Lawrence & Christensen 1979). Suppression of UV-sensitivity could still be a consequence of another DNA repair pathway being induced or enhanced in some way.

The *RAD*52 repair pathway probably repairs some UV-induced damage. The *SRS*2 mutation could not suppress *rad*6 UV-sensitivity by means of induction of the *RAD*52 pathway since *SRS*2 does not suppress

X-ray sensitivity in *rad*6 or *rad*18 mutants (Lawrence & Christensen 1979). The *SUQ*5 ochre suppressor, like *SRS*2, suppresses UV-sensitivity, but not UV-immutability in *rad*6-3 mutants; however, this suppressor has the opposite effect on the *rad*6-1 mutation, lending further support for the idea that there are separable error-free and error-prone repair processes under *RAD*6 control (Tuite & Cox 1981). The finding that *SRS*2 suppresses UV, but not X-ray, sensitivity is suggestive of at least two error-free *RAD*6-controlled repair processes. This is further supported by the finding that *rad*18 mutants are very UV-sensitive, but only slightly X-ray sensitive, while *rad*9 and *rad*15 mutants are very X-ray sensitive and only slightly UV-sensitive (Lawrence & Christensen 1979, Lawrence 1982).

Amongst the *RAD*6 group of genes clearly involved in mutagenesis, different genes are apparently concerned with the formation of different types of mutation, suggesting more than one *RAD*6-dependent error-prone pathway. The *REV*1 and *REV*2 genes enhance UV and X-ray induced reversion at some alleles only, with *REV*2 being highly specific for relatively few mutations. Using a series of *CYC*1 alleles as an assay system, it was shown that *REV*2 was required for reversion of the ochre allele *CYC*1-9, but not for a range of other mutations including ochre and amber alleles, initiation, misense, and frameshift mutations. In the same system, *REV*1 was involved in the UV-induced reversion of 10 out of 12 basepair substitutions and 3 out of 4 frameshift mutations. The *REV*3 gene was found to be involved in UV-induced reversion at most alleles tested including substitutions, deletions, and additions, although some specific alleles are apparently resistant to *REV*3-mediated reversion (Lemontt 1971, 1972, Lawrence & Christensen 1978a,b 1979). Different genes may also be involved in the generation of mutations by different agents: *RAD*6-dependent EMS-induced reversion (Prakash 1974) is independent of the *REV* genes and also of the *RAD*18 gene (Prakash 1976a,b).

In addition to these roles for *RAD*6 in a number of error-prone and error-free repair processes, this gene is apparently involved in some other aspect of metabolism since *rad*6-1 homozygous diploids are defective in sporulation (Cox & Parry 1968, Lawrence 1982). The biochemical nature of this defect is yet to be determined: premeiotic DNA synthesis occurs, but meiotic recombinants are not produced (Game *et al.* 1979), while spontaneous and induced mitotic recombination levels in *rad*6-1/*rad*6-1 mutants are enhanced (Hunnable & Cox 1971, Kern & Zimmerman 1978). Homozygous diploids for the *pso*1 mutation (Henriques & Moustacchi 1980) and for the *umr*2 and *umr*3 loci (Lemontt 1977) are also sporulation defective, but other loci in the *RAD*6 group: *RAD*5, *RAD*8, *RAD*9, *RAD*15, *RAD*18, *REV*1, *REV*3, *MMS*3, and *UMR*1, are apparently not involved (Cox & Parry 1968, Lemontt 1971, 1977, Prakash & Prakash 1977).

The involvement of the *CDC*8 gene in the *RAD*6 group is interesting because of the potential for this gene to influence mutation levels through its effect on nucleotide pools. In this way it has already been compared to mutants defective in deoxycytidylate kinase which show enhanced levels of mutation for some loci and reduced levels at other loci (Maus & Haynes

1984). It may also be significant that the induction of the error-prone SOS response in *E. coli* is accompanied by the induction of a unique deoxycytidylate deaminase (Estereron & Sicard 1986).

4.8.1 Recombinant DNA studies of the *RAD6* epistasis group

The *RAD6* gene has been cloned by complementation of *rad6* UV-sensitivity with a yeast genomic library or an extrachromosomal plasmid (Prakash *et al.* 1982), and independently by complementation of mms-sensitivity by Kupiec & Simchen (1984). Integrative deletion of the *RAD6* gene is not lethal to haploid cells, although these integrants do grow more slowly, suggesting that while the *RAD6* gene is not essential it may be involved in some aspect of cellular metabolism (Kupiec & Simchen 1984). Sequencing of the *RAD6* gene has predicted a protein of 19.7 kd which contains 23.3% acidic and 11.6% basic residues (Reynolds *et al.* 1985b). The protein sequence has revealed a unique region at the carboxyl terminal end in which 20 out of 23 amino acids are acidic. This sequence: Glu-Asp_2-Met-Asp_2-Met-Asp_{13}-Glu-Ala-Asp is reminiscent of the nucleosome-associated high-mobility group proteins HMG-1 and HMG-2, and thus may reflect the ability of *RAD6* protein to interact with nucleosomes, possibly having a role in making chromatin accessible to DNA repair proteins (Reynolds *et al.* 1985b).

4.9 THE *RAD52* EPISTASIS GROUP INVOLVED IN RECOMBINATIONAL PROCESSES

While the *RAD6* group is probably responsible for the majority of postreplicational repair after UV-irradiation, some component is dependent on the *RAD52* gene (Prakash 1981). In accordance with this, mutants in the *RAD52* group, although originally isolated because of their sensitivity to ionizing radiation (Resnick 1969, Game & Mortimer 1974), often show some slight cross-sensitivity to UV (Cox & Game 1974, Lawrence & Christensen 1976). The *RAD52* group of genes are all concerned with various aspects of gene conversion, recombination and DNA double-strand break repair (see below). However, replicational bypass of UV-induced DNA damage, as observed in mitosis and meiosis in *rad1* mutants, does not involve exchange of dimers between parental and newly-synthesized DNA (Resnick *et al.* 1983a,b).

The *RAD52* epistasis group includes the *RAD50*, *RAD51*, *RAD52*, *RAD53*, *RAD54*, *RAD55*, *RAD56*, *RAD57*, *RAD24*, and *CDC9* genes (Game & Mortimer 1974, McKee & Lawrence 1980, Haynes & Kunz 1981). Recombination has been studied extensively in yeast because it is ideally suited for this purpose. Yeast cells can grow continuously in haploid and diploid phase by mitotic division, and in addition diploids can be induced to undergo meiosis producing asci containing 4 spores, so that all 4 meiotic products can be scored. Recombination can be defined as the exchange of genetic information between homologous DNA duplexes. In yeast these exchanges can be interchromosomal, i.e. between homologous or non-homologous chromosomes which share a region of sequence homology, or intrachromosomal, i.e. between sister chromatids of the same chromosome, or between reiterated sequences on the same DNA duplex. In addition, recombination can occur between yeast chromosomes and gapped or linearized plasmids. Recombinational exchanges can be reciprocal or non-reciprocal; non-reciprocal exchanges (or gene conversions), are the results of information being transferred from one DNA duplex to another without concomitant transfer in the opposite direction. All of these recombinational processes are reviewed elsewhere with consideration of molecular models for recombination (Kunz & Haynes 1981, Dressler & Potter 1982, Orr-Weaver & Szostak 1985).

4.9.1 *RAD* genes and recombination

Several of the *RAD*52 group genes are involved in meiotic recombination (Game *et al.* 1980, Morrison & Hastings 1979, Prakash *et al.* 1980, Kunz & Haynes 1981). Since mutants in this group are very sensitive to ionizing radiation which induces double-strand breaks, and *rad*52 mutants are defective in repair of these breaks (Ho 1975, Resnick & Martin 1976, Mowat & Hastings 1979), it seems likely that a similar mechanism could be responsible for meiotic recombination and double-strand break repair. In accordance with this, haploid mutants from the *RAD*52 group do not exhibit the resistant tail on X-ray survival curves, which is typical of wild-type haploid cells. This tail is thought to reflect the relative resistance of G2 cells which presumably repair some DNA lesions by a mechanism dependent on a second copy of DNA.

While the *RAD*52 gene is required for all types of gene conversion, some mitotic reciprocal recombination and also intrachromosomal exchanges between reiterated genomic sequences on the same chromosome (Jackson & Fink 1981), or between sister chromatids (Prakash & Taillon-Miller 1981) are *RAD*52-independent. A more detailed examination of mitotic events does in fact suggest that at least some portion of mitotic recombination occurs by a separate mechanism altogether, or at least by a mechanism that has some different steps. The requirement of *RAD*52 for all mieotic gene conversion, but not mitotic recombination, indicates that these two events are separable, and this is in accordance with the finding that mitotic gene conversion is less closely associated with reciprocal recombinations. Although UV, nitrosoguanidine, and other DNA-damaging agents enhance both events (Fogel & Hurst 1963), UV enhances reciprocal recombination more than gene conversion, while nitrosoguanidine enhances gene conversion preferentially (Roman & Fabre 1983). Furthermore, different mutant alleles have differing effects on these processes. The *spo*ll mutation, like *rad*52, abolishes meiotic recombination, but only decreases mitotic recombination (Prakash *et al.* 1980). The *rad*50 mutation, however, which is required for meiotic recombination, shows elevated levels of spontaneous, but not radiation-induced, mitotic gene conversion or reciprocal recombination (Malone & Esposito 1981). The *rem*1-1 mutation has also been examined, and enhances mitotic gene conversion and reciprocal recombination, but has no effect on meiotic processes (Golin & Esposito 1977, Malone *et al.* 1980).

Mutations in the *RAD*6 or *RAD*18 genes can also lead to enhanced levels of mitotic recombination and gene conversion, and this may be a consequence of the impairment of the repair processes in this group, so that more lesions are channeled into the *RAD*52 pathway. The *RAD*3 gene is intriguing in this respect: while other mutants in the *RAD*3 group show increased levels of radiation-induced mitotic gene conversion and reciprocal recombination, *rad*3 mutants show elevated spontaneous levels of these events, and also higher levels of UV-induced mitotic recombination, but lower levels of UV-induced gene conversion (Kern & Zimmerman 1978).

Analysis of mitotic recombination events in yeast as described above have led to the idea that the mechanisms for these events are different from those in meiosis. Meiotic recombination occurs after DNA replication when there are four DNA duplexes; most mitotic gene conversion apparently occurs during G1, i.e. at the second strand stage (Esposito 1978, Fabre 1978). Mitotic gene conversion is also more frequently associated with the formation of symmetric heteroduplex DNA (Golin & Esposito 1981), and the polarity observed in meiosis is abolished at the *his*1 gene in mitosis (Hurst & Fogel 1964). In addition, meiotic recombination is essential since in the absence of recombination, meiotic division leads to lethality (Game 1984). Recombination during mitosis, however, may just be a consequence of, or a mechanism for, replication past DNA damage. Mitotic recombination also seems to reflect the operation of a *RAD*52-dependent pathway responsible for gene conversion and some reciprocal recombinations, probably those associated with gene conversion, and a *RAD*52-independent pathway for reciprocal recombination (Orr-Weaver & Szostak 1985).

4.9.2 Recombinant DNA studies on the *RAD*52 epistasis group

To date the *RAD*50, *RAD*51, *RAD*52, *RAD*54, *RAD*55, and *RAD*57 genes have been cloned by complementation of the ionizing radiation or MMS-sensitivity of the corresponding mutants (Calderon *et al.* 1982, 1983, Schild *et al.* 1983, 1982 Adzuma *et al.* 1984, Kupiec & Simchen 1984b). Disruption of the *RAD*54 gene does not result in lethality, indicating that this gene is not essential (Calderon *et al.* 1982, 1983). DNA sequence analysis of the *RAD*52-1-complementing fragment has indicated an open reading frame of 1512 nucleotides, and S1 mapping suggests formation of an mRNA which is not spliced (Adzuma *et al.* 1984).

As yet, none of the cloned genes in the *RAD*52 group have been reported to have been over-expressed and the proteins isolated. There is, however, evidence that the *RAD*52 gene controls a deoxyribonuclease with maximum activity during early premeiotic DNA synthesis and commitment to recombination (Chow & Resnick 1983)

4.10 THE ROLE OF YEAST DNA REPAIR PROCESSES IN CHEMICALLY-INDUCED DNA DAMAGE

So far we have confined our consideration of yeast DNA repair to those processes involved in the repair of radiation-induced DNA damage. We will now consider the roles for these processes and perhaps other unique processes in the repair of chemically-induced damage. With respect to the latter, many yeast mutants have been isolated by sensitivity to chemical agents: *snm* mutants sensitive to nitrogen mustard (Ruhland *et al.* 1981), *pso* mutants sensitive to photoactivated psoralens (Cassier *et al.* 1980, Henriques & Moustacchi 1980), and *mms* mutants to methylmethanesulphonate (Prakash & Prakash 1977), and more recently a mutant sensitive to *N*-methyl-*N'*-nitro-*N*-nitrosoguanidine ((MNNG): *ngs*1 (Nisson & Lawrence 1986). Some of these mutants are also sensitive to radiation, and similarly the radiation-sensitive mutants show cross-sensitivity to chemical agents.

4.10.1 Excision processes

The involvement of the *RAD*3 group in the repair of UV-induced damage and the cross-sensitivity of mutants in this group to polyfunctional alkylating agents (Brendel & Haynes 1973, Ruhland & Brendel 1979), mitomycin C (J. C. Game, cited in Game & Cox 1972), mono- and bifunctional psoralens (Averbeck & Moustacchi 1975, Averbeck *et al.* 1978), but not to ionizing radiation, parallels the situation in *E. coli.* In *E. coli,* the *UVR*ABC system is responsible for the removal of bulky adducts which cause helical distortion (see Friedberg 1985a for a review). Minor base damage caused by simple monofunctional alkylating agents, on the other hand, is repaired by a system of DNA glycosylases, apurinic/apyridimic (AP) endonucleases, exonucleases, DNA polymerases, and ligases; some specific damage is removed by methyltransferase activity (Lindahl 1979, 1982, Lindahl *et al.* 1983, McCarthy *et al.* 1983, 1984).

In accordance with this, *uvr*A mutants are not sensitive to MNU (Hince & Neale 1974), and O^6methylguanine is not repaired by the *UVR*ABC pathway (Warren & Lawley 1980). There is evidence, however, that the *UVR*ABC system can repair alkylation damage for groups larger than methyl (Sedgwick & Lindrahl 1982, Friedberg 1985a). In yeast, no methyltransferases or DNA glycosylases for alkylation damage have been identified to date. However, wild-type yeast cells possess AP endonucleases (Armel & Wallace 1978, 1984, Akhmedor *et al.* 1982, Foury 1982), and wild-type cell extracts can remove 3-methyladenine and 7-methylguanine from alkylated DNA (Nilsson & Lawrence 1986). In addition, all ENU-induced

DNA single-strand breaks are repaired in yeast cells, suggesting that yeast may also be able to repair ethylphosphotriesters (Cooper & Waters — unpublished observations). This is an interesting observation since while *E. coli* repair alkylphosphotriesters by means of an inducible methyltransferase (McCarthy *et al.* 1983), mammalian cells do not repair these adducts (Shooter & Merrifield 1976, Shooter & Slade 1977, Shooter *et al.* 1977, Warren *et al.* 1979, Bodell *et al.* 1979).

*rad*1, *rad*2, *rad*3, *rad*4, *rad*10, *rad*14, and *rad*16 mutants are not EMS-sensitive (Lawrence *et al.* 1974, A. J. Cooper & R. Waters — in preparation), and *rad*1 *rad*2 and *ngs*1 *rad*1 *rad*2 mutants can remove 3-methyladenine and 7-methylguanine from alkylated DNA (Nisson & Lawrence 1986). In addition, a range of mutants have been isolated by sensitivity to MMS (Prakash & Prakash 1977), and many of these do not show cross-sensitivity to radiation, suggesting that there may be a unique pathway (or pathways) for alkylation damage in yeast. However, several of the *mms* mutants are sensitive to UV or X-rays or both (Prakash & Prakash 1977). In fact one of the *mms* mutants, *mms*19, seems to be in the *RAD*3 group, and this mutant has been shown to be defective in dimer removal in specific biochemical studies (Prakash & Prakash 1979). In addition, *rad*1 and *rad*4 mutants have been reported to be MMS-sensitive (Prakash & Prakash 1977) although they are not sensitive to MNU (Cooper & Waters — in preparation). This difference may be due to the difference in the pattern of DNA alkylations obtained with the two agents. In addition to *rad*1-1 and *rad*4-4, *rad*2-1, *rad*3-2, *rad*10-1, and *rad*14-2 were found to give wild-type responses to MNU, implying that generally the excision repair genes are not necessary for the repair of methylations (Cooper & Waters — in preparation). However, although none of the *rad*3 group of mutants have been found to be EMS-sensitive, *rad*1-1, *rad*2-1, *rad*4-4, and *rad*14-2 are ENU-and ENNG-sensitive, suggesting that they are involved in the repair of some DNA ethylations. In the same study *RAD*3-2 and *rad*10-1 mutants gave a wild-type response to these, agents suggesting that these genes may be exclusive to the repair of bulky adducts in DNA (Cooper & Waters — in preparation). Complicating the issue even further, *rad*1-1 and *rad*2-1 mutants were also found to be sensitive to DES, but *rad*3-2, *rad*4-4, *rad*10-1, *rad*14-2, and *rad*16-1 mutants showed no increased sensitivity to this agent. Thus it seems that mutants in the *RAD*3 group can be divided into at least three groups for the repair of alkylations. This is a surprising finding, since the *RAD*3 epistasis group is well defined in as much as virtually all of the mutants in this group have been shown to be deficient in pyrimidine dimer removal by biochemical studies (Prakash 1977a,b, Prakash & Prakash 1979, Reynolds & Friedberg 1981, Miller *et al.* 1982). These studies, together with the finding that these mutants show cross-sensitivity to chemical agents which cause bulky adducts, has led to the idea that all of these gene products act in one pathway, possibly in the form of an excision repair complex analogous to the *UVR*ABC complex to excise DNA damage. While this may all still be true as far as some adducts are concerned, it now seems that these various *RAD* proteins may act in a number of different repair pathways. This may possibly be through the formation of different repair complexes which may or may not involve proteins from the other epistasis groups, to repair different types of DNA damage (Cooper & Waters — in preparation).

4.10.2 Misrepair and recombination processes

The *RAD*6 epistasis group has already been shown to reflect a number of different repair processes, and accordingly there are differences among alkylation sensitivities here too (Haynes & Kunz 1981, Lawrence 1982, Cooper & Waters — in preparation). Mutants in the *RAD*52 group also show cross-sensitivity to photoactivated psoralens (Henriques & Moustacchi 1980) and to alkylating agents (Haynes & Kunz 1981, Cooper & Waters — in preparation), suggesting that some chemical damage may also be repaired by a recombinational mechanism. Interestingly, EMS-induced reversion of *CYC*1-131 and *CYC*1-115 is reduced in *rad*50, *rad*51, *rad*52, *rad*54, and *rad*56 stationary phase mutants, indicating that these genes may play a role in the generation of EMS-induced mutations while the role of these genes in radiation repair is apparently error-free (Prakash 1976a, Prakash & Higgins 1982). It has been suggested that some EMS-induced damage is repaired by an error-prone repair pathway involving these genes from the *RAD*52 group and the *RAD*6 gene (Prakash & Higgins 1982). This is supported by the finding that *rad*6-3 *rad*52-1 double mutants are no more EMS-sensitive than the most sensitive single mutant. However, in the same study *rad*6-1 *rad*52-1 mutants were found to be considerably more sensitive than either of the single mutants (Prakash & Higgins 1982). Thus it seems that double mutant studies using different alleles can give varying results, and this is important when considering the validity of the epistasis groups as they stand. Similarly, as in the case of the *RAD*7 and *RAD*23 genes, if two genes can complete the same repair step, they behave synergistically although they may still be in the same repair pathway. The data available also imply that the examination of double mutants for sensitivity to different DNA damaging agents may place the *RAD* mutants in different epistasis groups (Cooper & Waters — in preparation).

It is interesting that to date no O^6alkylguanine-DNA methyltransferase has been identified in yeast cells, since this enzyme has been found in bacteria (Robins & Cairns, 1979, Foote *et al.* 1980, Olsson & Lindahl 1980) and mammalian cells (Mehta *et al.* 1981, Pegg & Perry 1981, Hans *et al.* 1983, Krokan *et al.* 1983). It is still of course possible that a methyltransferase exists in yeast but has not been isolated. Yeast cells, however, do not exhibit an adaptive response to alkylation, as is observed in *E. coli* (Samson & Cairns 1977) *Baccillus subtilis* (Haddon *et al.* 1983, Morohosh & Munakata 1983) and *Micrococcus luteus* (Ather *et al.* 1984), although *DIN*1, *DDR*48, and *DDRA*2 genes are MNNG inducible (Maga & McEntee 1985). An inducible adaptation response is not universal since it is apparently absent from *Salmonella typhimurium* (Ather *et al.* 1984) and *Haemophilus influenzae* (Kimball 1980), and the methyltransferase may not be inducible in mammalian cells (Foote & Mitra 1984, Montesano *et al.* 1979, 1980, Lindamood *et al.* 1983, Kleinies & Margison 1976, Pegg 1978,

Charlesworth *et al.* 1981, Pegg & Perry, 1980). Similarly, no DNA glycosylases for alkylation damage have been identified; certainly the *ngs*1 mutant does not seem to reflect a deficiency in removal of 3-methyladenine or 7-methylguanine, at least by means of a glycosylase activity, since these lesions are repaired in this mutant (Nisson & Lawrence 1986). A base excision repair mechanism for alkylation repair cannot be ruled out, however, as most of the alkylation-sensitive mutants have not as yet been characterized, and the presence of AP endonucleases in yeast is suggestive of base excision repair mechanisms. The sensitivity of some of the excision repair *RAD* mutants could reflect common steps between these two types of excision repair. It does, however, seem likely that yeast cells possess a unique repair mechanism for alkylation damage.

4.11 ASPECTS OF MUTAGENESIS IN YEAST (See Chapter 6)

Yeasts provide a convenient organism for eukaryote comparison to mutagenesis data from *E. coli* and to ask questions which are of unique relevance to eukaryotes. The mechanisms of mutagenesis which have been mentioned previously have been described in terms of changes occurring in DNA during DNA replication or DNA repair. Of particular interest is the requirement in *Saccharomyces cerevisiae* for a functional *RAD*6 gene product for almost all induced mutation.

4.11.1 General background

Mutagenesis studies in *Saccharomyces cerevisiae* have relied heavily on genetic analysis to probe the nature of mutations and processes effecting mutagenesis. As described in section 4.8, the *RAD*6 Epistasis Group is involved in mutagenesis, but has a function in error-free and error-prone repair (Lawrence & Christiansen 1979, Tuite & Cox 1981). The *RAD*6 gene product is required for practically all induced mutation with limited known exceptions such as the reversion of the frameshift allele *his*4-519 by ICR 170 (Walsh & Fink, cited by Lawrence 1982).

Studies on the *RAD*3 and *RAD*52 Epistasis Groups have indicated that the radiation-damage repair undertaken by these pathways is error-free. The evidence for this includes the approximately equal mutation levels at equal UV survival levels found in wild-type *RAD* and mutant *rad* strains of the excision pathway (Eckardt & Haynes 1977), and the hypermutability of strains carrying *rad*52 *rad*1 double mutations suggesting channelling of lesions into error-prone repair. With regard to the repair of ionizing radiation DNA damage, where — unlike for UV damage — the *RAD*52 Group has the major repair contribution, then again *RAD*6 appears to represent the sole pathway for mutagenesis.

Prakash (1974, 1976a,b) has examined the requirement of yeast for a functional *RAD*6 gene product in chemical induced mutation, and has found a dependence on *RAD*6 function. This occurred with agents which are classically considered to induce bulky lesions subject to DNA repair, e.g. 4 nitroquinoline-1-oxide (4NQO), as well as with agents like *N*-methyl-

N^1-nitro-*N*-nitrosoguanidine (MNNG) which produce lesions which give rise to mispairing during DNA replication. The later finding of reduced mutagenesis at *cyc*1-131 and *cyc*1-115 after ethylmethane-sulphonate treatment of *rad*50, *rad*51, *rad*52, *rad*54, and *rad*56 stationary phase cells indicates that there may be a role for these genes in misrepair of EMS induced damage (Prakash & Higgins 1982). The role of the *RAD*6 pathway in mutagenesis, but distinct from recombination, is a distinction from the *E. coli* data and with regard to the requirement of *RAD*6 for mutagenesis by chemicals which classically induce mispairing-type damage. In *E. coli* these forms of potential mispairing lesions are not repaired by the error-prone mechanism controlled by *REC*A and *LEX*A (for review, see Friedberg 1985).

In *E. coli* the mutant *umu*C has been isolated which effects induced mutagenesis as part of the cascade of 'SOS' functions regulated by *REX*A and *LEX*A (Witkin 1974, Kato & Shinoura 1977). Earlier, mutants of yeast had been isolated which exhibited reduced mutagenesis (Lemontt 1971) and assigned to three loci *REV*1, *REV*2, and *REV*3 where *REV*2 was found to be allelic with *RAD*5 (Game & Cox 1971). These loci were isolated on the basis of a reduced UV-induced mutation phenotype using base-substitution markers. The *REV* genes appear to be involved in mutagenesis in differing ways, with the *rev*3 mutant effecting UV mutagenesis at a variety of genetic sites, *rev*2 effecting some ochre allele reversions, and *rev*1 effecting many transition and transversion mutagenesis events (Lawrence & Christensen 1978a,b, 1979). Other studies of mutagenesis in these strains after treatment with different DNA-damaging agents than UV indicate the complexity in analysis of such effects as the apparent independence of the *REV* genes from involvement in EMS-induced reversions (Prakash 1974). This and other findings are discussed in section 4.8, indicating that misrepair may not represent a single pathway following the *RAD*6-mediated step.

Other mutants have been isolated which effect mutagenesis and include the *umr* mutations which were isolated on the basis of reduced UV-induced forward mutation to canavanine resistance (Lemontt 1977). Seven genetic loci were defined for the *umr* mutants which were unlinked to the structural gene for arginine permease, which is the altered gene product in canavanine-resistant mutants. Of the *umr* mutants, *umr*1, *umr*2, and *umr*3 exhibited the characteristics of slight UV sensitivity and an absence of general effects on cell physiology and morphology. These genes may encode products for specific functions for specific types of mutations, as their effect on reversion of three different ochre alleles was variable.

Analysis of the MMS-sensitive mutant *mms*3 revealed a new gene involved in mutagenesis with *a*/α *mms*3/*mms*3 diploids and *mms*3 haploids exhibiting defective mutagenesis after UV treatment (Martin *et al.* 1981). The interesting result was obtained that *mms*3 diploids which were homozygous at the mating-type locus, i.e. a/a or a/α were normal in UV reversion.

The selection methods employed in isolating mutants involved in mutagenesis have been critized for being limited in the type of mutation studied after specific types of DNA damage (Lawrence 1982). Using a different selection scheme, Lawrence *et al.* (1985) have isolated a new gene concerned with UV mutagenesis designated *REV*7. They used a method which included the use of UV, MNNG, and MMS to produce a variety of DNA damage, and their effects on nonsense, missense, and frameshift reversion were included. Genes effected in frameshift reversion have also been isolated using the *his*4-38 allele (cited by Lawrence *et al.* 1985), and this methodology coupled to recombinant DNA techniques should enable analysis of mutagenesis in biochemical detail.

4.11.2 Spontaneous mutation

In addition to the mutants effected in induced mutagenesis which are outlined above, mutants have been derived which are effected in spontaneous mutation, and the spontaneous mutator phenotypes of other strains effected in DNA repair processes have been examined. It has been suggested from the alteration of spontaneous mutability in DNA repair defective strains that 90% or more of spontaneous mutation in *S. cerevisiae* is due to mutagenic repair (Hastings *et al.* 1976, Quah *et al.* 1980).

Mutator genes have been isolated which exert an effect which enhances spontaneous mutation to various degrees for various mutation markers (von Borstel *et al.* 1973, Hastings *et al.* 1976, Gottleib & von Borstel 1976). Among these mutants *mut*1-2 and *mut*2-1 have been found to yield an additive effect, except in double mutant diploids, which may indicate alternative pathways of spontaneous mutagenesis (Gottlieb & von Borstel 1976). These studies have been developed through the isolation of the *ant*1 mutant which partially reverses the *mut*1-1 phenotype, and *ant*2 which is allelic with *REV*3 (Quah *et al.* 1980). The finding that some DNA repair mutants isolated in studies on induced damage repair have an opposite effect, or no effect on spontaneous mutagenesis, in comparison to induced mutagenesis, has been interpreted in terms of lesion channelling where separate pathways to the *RAD*6 pathway may exist (Hastings *et al.* 1976, Quah *et al.* 1980).

The scale of the effects of mutators and anti-mutators and the mutations used in analysis of their effect has been called into question by Lawrence (1982) who pointed out that these small affects may be due to indirect or peripheral effects on spontaneous mutagenesis. The effect of the mutators and anti-mutators most markedly or specifically at suppressor loci may also lead to difficulties in interpretation as they may be concerned with aspects of the selection of expression of the mutations.

Other mutator genes have been described which include the mutant *rem*1 which enhances mitotic recombination and forward mutation to canavanine resistance and for tryptophan independence at the *trp*5-12 locus (Golin & Esposito 1977). The *DEL*1 mutator gene has also been found to exert a mutator effect on the adjacent *CYC*1, *RAD*7, and *OSM*1 gene cluster in a mechanism which may involve transposable elements (Liebman & Downs 1980), and deletions at the *SUP*4 locus may have a similar origin (Rothstein 1979).

In addition to genes exerting an effect on nuclear mutagenesis, mutants have been isolated which exert an effect on spontaneous mitochondrial mutagenesis. These include *gam*2 and *gam*4 (Foury & Goffean 1979) and *mtm*1 and *mtm*2 (Johnston 1979, Johnston & Johnston 1983) which are nuclear genes which enhance *petite* induction and mitochondrial point mutations, but not the nuclear gene mutations examined. The mutator gene *mtm*3 was found to enhance the frequency of various mitochondrial and nuclear gene mutations, but not *petite* mutagenesis (Johnston & Johnston 1983), while the anti-mutators LB_6 reduced the frequency of a variety of nuclear and mitochondrial mutations (Bianchi & Foury 1982). These findings as a whole would appear to imply that some prerequisites for spontaneous mutation are in common for nuclear and mitochondrial genes, but that there are some unique requirements for mitochondrial mutagenesis.

The nature of mutations occurring during spontaneous mutagenesis have been examined by Sherman *et al.* (1974) where more than 80% were attributed to base-substitution using the iso-1-cytochrome C gene. Whelan *et al.* (1979) support this finding, and among 233 forward mutants to canavanine resistance may not have detected any frameshift or large deletions. Thus it is likely that base substitution mutation is the most common spontaneous gene mutation event.

Of particular interest in studies on spontaneous mutation in *S. cerevisiae* is the 'meiotic effect' of enhanced frameshift mutation during meiosis (Magni & Puglisi 1966). This was originally observed as an increase in the reversion rates at certain loci in meiosis (Magni & von Borstel 1962), and a similar effect for spontaneous frameshift mutation in *Schizosaccharomyces pombe* has been reported (Friis *et al.* 1970). Magni & Puglisi (1966) suggested that the enhanced frameshift mutation rate was associated with the recombination process as it was associated with outside markers exchange. It may be that these effects result from single-strand gaps in the recombination process or as a result of the incomplete meiotic DNA synthesis prior to minor DNA synthesis in meiotic prophase I which has been detected in *Lilium* (Stern & Hotta 1973).

Mutation frequences and types have been examined for forward mutation to canavanine resistance in meiosis and mitosis by Whelan *et al.* (1979) and Gocke & Manney (1979). No meiotic effect was demonstrated, but this may have been related to an absence of the frameshift class. Lawrence (1982) has commented on these results, and has stressed the importance of complete detection of mutations which may increase estimates of mitotic rates of mutation. Other frameshift mutation studies have used the *his*4-519 allele which has been stated to exhibit a 'meiotic effect' (Panzeri *et al.* 1979) but which fails to give an increased spontaneous reversion frequency in meiosis in other studies (Machida & Nakai 1980, Kelly 1982). The 'meiotic effect' was also found to include ochre suppressor type mutations in the original studies (Magni & Puglisi 1966), and these have been attributed to base–pair substitutions (Piper *et al.* 1976). This indicates that if the 'meiotic effect' for spontaneous mutation exists, it also includes some other mutation classes besides frameshift point mutations (Lawrence 1982).

4.11.3 Mitochondrial mutagenesis (see chapter 2 section 6.1)

In addition to the mutants described above, mitochondrial mutagenesis has been examined through the isolation of mutants which are sensitive to UV-induced *petite* induction (Moustacchi *et al.* 1976). Of the five mutants isolated which were designated *uvsρ* three represented nuclear mutations, and other *rad* mutants have been found to influence *petite* induction (Moustacchi 1969, 1971). The *gam* strains isolated by Foury & Goffeau (1979) and the *mtm* strains isolated by Johnston (1979) have been examined for their effect on spontaneous mutation, and they contribute to the evidence for distinct gene products involved in mitochondrial mutagenesis.

Other agents which are mutagenic for nuclear genes also act on the mitochondrial genome (Ejchart & Putrament 1979, Polakowska & Putrament 1979, Baranowska & Putrament 1979). Treatments which effect mitochondrial mutagenesis exclusively also exist, such as thymidylate deprivation (Barclay & Little 1978), while other intercalating agents, in particular ethidium bromide, are particularly effective at inducing *petite* mutations in up to 100% of the cell population (Slominski *et al.* 1968). Other agents may induce *petite* mutations through non-genotoxic action such as the sterol synthesis inhibitor ketoconazole (Bligh & Kelly, in preparation); such effects should be remembered in evaluating *petite* induction data.

4.11.4 Inducibility of mutagenesis: (see chapter 3)

Studies on mutagenesis in *E. coli* have revealed the error-prone 'SOS' repair mechanism which is inducible by damage which interferes with DNA synthesis (for reviews, see Friedberg 1985a). These studies have been interpreted to reflect the development of a system which can repair DNA damage, albeit with errors, producing reduced cell lethality (Radman 1974). However, the mutant *umu*C only contributes a slight effect to UV resistance to lethality, but abolishes UV mutagenesis (Bagg *et al.* 1981), and so the mutagenesis component may be for generating variability.

Evidence for the inducibility of mutagenesis in *S. cerevisiae* has been examined. Eckardt and Haynes have investigated aspects of the kinetics of gene reversion induction (Haynes & Eckardt, 1979a,b, review by Siede & Eckardt 1984). Biphasic and linear-quadratic kinetics for reversion were found in wild-type cells where the quadratic component was taken to reflect inducible functions and the linear component constitutive processes. In contrast to the wild-type response, excision-defective strains exhibited biphasic and purely linear kinetics for reversion, while no quadratic component was found for forward mutation induction in repair proficient or deficient strains (Eckardt & Haynes 1977, Kilbey *et al.* 1978). Support for the inducible functions reflected in the quadratic component was obtained using treatment with cyclohexamide which was found to shift UV reversion induction kinetics from linear-quadratic to near linear (Eckardt *et al.* 1978).

Split-dose experiments with a liquid holding period between treatments have also been used to examine the inducibility of UV mutagenesis (Patrick & Haynes 1968, Eckardt *et al.* 1978). The treatment of yeast with an initial

UV dose followed by liquid-holding prior to the second UV dose resulted in increased mutation and survival, but this was abolished by treatment with cycloheximide during liquid holding which is consistent with the need for synthesis of new proteins to give increased mutability and survival (Eckardt *et al.* 1978).

Postreplication repair is *RAD*6-dependent in yeast, and Pakash (1981) found it to be sensitive to cycloheximide, thus implicating the requirement for protein synthesis. This component of *RAD*6 repair is likely, however, to be error-free (Lawrence 1982) and not to contribute to mutagenesis. The use of a *rev*2 temperature sensitivity mutant has been used by Siede *et al.* (1983a,b) to probe directly inducibility of mutagenesis of the type found in this mutant. As the *rev*2 defect is restricted to reversion of certain types of ochre alleles the overall relevance of these findings to mutagenesis in general has been called into question (Siede & Eckardt 1984). Siede *et al.* (1983a) found for *REV*2 activity a sensitivity to cycloheximide providing evidence for inducibility at least for this reversion type in stationary phase cells. In contrast to the total inhibition of *REV*2 activity by cycloheximide in stationary phase cells it was only reduced in exponential phase cells, which may reflect a higher constitutive level in growing cells (Siede *et al.* 1983b).

Evidence of inducible functions in mutagenesis has been investigated using mating experiments between irradiated strains carrying a *cyc*1-363 nonrevertible deletion mutation and an unirradiated strain carrying a *cyc*1-91 point mutation. In this way it was possible to study whether mutations could be induced in DNA which had not received DNA damage (Lawrence & Christensen 1982, Lawrence *et al.* 1982). The level of untargeted mutagenesis was found to be unlikely to be less than 40% of the mutations normally generated.

The phenomenon of untargeted mutagenesis has also been used to detect whether cytoplasmic factors are involved by using the *kar*1 mutation to prevent nuclear fusion during mating. This was found to prevent untargeted mutagenesis, which argues against the induction of cytoplasmic proteins which can produce untargeted mutations, but the results overall may apply only to those special conditions. These conditions included the use of excision-deficient cells in the experiments because mutations are fixed within the period prior to mating in wild-type cells, and it has been noted by Siede & Eckardt (1984) that such excision-deficient strains do not exhibit quadratic reversion kinetics associated with inducible processes.

Other studies on the inducibility of mutagenesis have examined the activity of proteases in yeast. The rationale for this approach is based on the role of the *REC*A protease of *E. coli* which is induced and cleaves the repressor protein *LEX*A of the 'SOS' genes. Proteinase B activity was found to be induced in a *RAD*6-dependent manner in response to UV damage (Schwencke & Moustacchi 1982a,b), but the biological significance of these studies remains unclear.

Molecular biology is likely to provide the firmest and more precise evidence for inducibility of components of the mutagenesis process by gene fusion and transcription studies. The identification of damage-inducible genes was outlined in section 4.7, and it is possible that some of these may represent aspects of error-prone repair. The cloning of genes which have specific effects on mutagenesis, e.g. *REV*3, may also permit these questions to be addressed.

4.11.5 Specificity of mutagenesis (see chapter 6)

Examination and analysis of the specific types of mutations found spontaneously and after treatment with mutagenic agents has been made, using test systems indicative of particular base changes. These have included analysis of the iso-1-cytochrome C gene (Prakash & Sherman 1973, Lawrence & Christensen 1979) and the *his*4 gene (Donahue *et al.* 1980).

Specificity of mutation has been found in other systems and has been studied in most detail in *E. coli*. 'Hot-spots' of mutation have been found as well as particular types of mutations dependent on the mutagenic agents used. Prakash & Sherman (1973) examined twelve mutagens for activity with eleven test *cyc*1 alleles, and found that five including MMS, DMS, HN_2, UV, and γ-rays gave approximately equal reversion at different alleles. The remaining mutagens, which were EMS, DES, NG, NIL, NA, tritiated uridine, and β-propiolactone, all reverted the *cyc*1-131 allele to a significantly greater extent. The difference between this allele and others examined was in the possibility of GC-AT transitions as a mechanism of reversion. This was taken as evidence that either the mutagens were specific or preferential for GC pairs, or that this was a 'hot spot' for DNA damage of the type induced. Selective action for 4NQO has, however, been shown for GC base pairs (Prakash *et al.* 1974, Prakash & Sherman 1974). In addition it appears that the *cyc*1-6 allele which can revert by mutation at a GC pair is not sensitive to 4NQO which has been taken to indicate it to represent a 'cold spot' for mutation.

The mutagens outlined above appear to induce base-pair substitution primarily with much lower induction of frameshift mutation. The spectrum of mutations induced by γ-rays included most frameshift mutations with approximately 20% in this class (McKee & Lawrence 1979a,b). More specific capacity for frameshift induction has been found for the acridine half-mustard ICR-170 (Culbertson *et al.* 1977, Donahue *et al.* 1980) and for hycanthone (Lucchini *et al.* 1980).

The data on variations in mutagenesis dependent on the mutagen and the site of mutations is of interest in *S. cerevisiae* because of the requirement for *RAD*6 in mutagenesis. The cause of the variations in terms of the targeting of mutations to sites of DNA damage has been called into question by Lawrence (1982) who found that in excision-deficient strains high levels of untargeted mutation occur, at least after UV treatment. Other mutagens may give rise to preferential sites for mutation by inducing certain lesions e.g. 4NQO or ICR-170, but the cause of the variation in reversion of identical base-pair changes at different sites in the gene remains uncertain.

For UV treatment it appeared not to be influenced by nucleotide environment, and may be due instead to overall chromatin structure (Lawrence 1982).

4.11.6 Genetic effects and the cell cycle: (see chapter 6 section 4.4.5.2)

Saccharomyces cerevisiae offers a convenient organism to probe the sensitivity of cells at different points in the cell cycle, to compare sensitivities of cells at different parts of the growth curve, i.e. the stationary and exponential phases, as well as to compare sensitivities of cells in meiosis and mitosis. In these studies the term 'cell age' has been applied in some cases which encompasses development through cell cycles or growth phases (Parry *et al.* 1976).

Early studies on these phenomena were made by Beam *et al.* (1954) who observed that the budding cell fraction exhibited radiation resistance. A general rule has been found to apply that for radiation damage growing cells are more resistant than stationary phase cells, while for chemical mutagens growing cells are more sensitive than stationary phase cells. Those differences have been predominantly ascribed to DNA repair variations, although variations in permeability for chemical agents and of access to the DNA with DNA replication and changes in chromatin structure may represent other causes.

Studies on ionising radiations have been developed to include studies on their effect on cells in the different stages of the mitotic cell cycle i.e. G1, S, G2/M phases. Maximum resistance has been found associated with and after the period of DNA synthesis in S/G2 phase cells (Bacchetti *et al.*, 1967; Fabre, 1973; Hatzfeld and Williamson, 1974; Brunborg and Williamson, 1978). This variation from G1 phase to S/G2 has been inferred to be due to differing repair capacities from the absence of relative resistance in repair deficient strains (Game & Mortimer, 1974; Rao & Reddy, 1982). The biochemical analysis of the nature of the repair of double-strand DNA breaks has revealed that G1 phase cells are unable to repair such lesions unlike G2 phase cells (Brunborg *et al.*, 1980).

The repair mechanisms(s) implicated in the repair double strand breaks are recombinational (Resnick, 1976) and require a duplicate genome, but as diploid yeast strains still exhibit such cyclic variation and have a duplicate genomethen Brunborg & Williamson (1978) attributed the S/G2 resistance to a sister-chromatid exchange repair mechanism. The sensitivity of cells in G1 phase have been associated with enhanced mutation frequencies (Tippins, 1978) which may implicate the lower capacity of cells in G1 phase to repair such ionising radiation damage by error-free sister-chromatid recombination repair.

Similar fluctuations in sensitivity and mutation induction between G1 and S/G2 phase cells have been found after UV irradiation of synchronised yeast cultures (Esposito, 1968; Parry & Cox, 1968; Chanet *et al.*, 1973;

Davies *et al.*, 1978). The period of greatest cell sensitivity to the lethal effects of UV was observed in the G1 phase of growth which has been attributed to recominational repair (Fabre, 1973) and G1 phase cells exhibited maximum induction of genetic end points. The possibility that the variations observed were due to different quantities of UV damage in the form of pyrimidine dimers was examined by Chanet *et al.* (1974) who observed 30% less dimers in S phase cells, which is unlikely to explain the bulk of the observed variation. Using a recombination defective strain recombination repair processes were implicated as the main cause of S/G2 resistance Chanet *et al.* (1974).

Sensitivity to chemical mutagens has also been examined in relation to the mitotic cell cycle with 'S' phase cells appearing most susceptible to lethal effects, with increased frequency of induced mutation. Data on *N*-methyl-*N′*-nitro-*N*-nitrosoguanidine (MNNG) is most comprehensive and has been used because of the preferential mutagenesis observed during DNA replication in bacteria (Cerda-Olmedo *et al.* 1968). In yeast MNNG has been found to give enhanced gene mutation during 'S' phase, and slight temporal variation was taken as evidence of the timing of replication of individual genes (Kee & Haber 1975, Burke & Fangman 1975). MNNG has also been used as a probe to investigate the timing of mitochondrial DNA replication (Dawes & Carter 1974).

The cause of such replication point sensitivity has been attributed in bacterial studies to direct mispairing with little or less time available for the repair of premutagenic lesions (Sklar & Strauss 1980), but may be due to misincorporation of alkylated bases from the nucleotide pools at this time (Topal *et al.* 1982). In *Saccharomyces cerevisiae* then, similar rationale may be applied, but the requirement for *RAD*6 for mutagenesis in stationary phase cells has not been extended to such cell-cycle studies.

The mutagen MMS has been found to induce double-strand breaks (Chlebowicz & Jachymczyk 1979) and MMS-induced damage has similar repair requirements to ionizing radiation-induced damage (Prakash & Prakash 1977). Increasing sensitivity was reported by Chlebowicz & Jachymczyk (1979) for cells in G1 phase arrest which were unable to repair MMS-induced double-strand breaks. These cells are more sensitive than stationary phase cells. The differences observed have been predominately ascribed to DNA repair variations, although variations in permeability for chemical agents and of access to the DNA with DNA replication and changes in chromatin structure may represent other causes.

Studies on ionizing radiations have been developed to include studies on their effect on cells in the different stages of the mitotic cell cycle, i.e. G1, S, G2/M phases. Maximum resistance has been found associated with and after the period of DNA synthesis in S/G2 phase cells (Bacchetti *et al.* 1967, Fabre 1973, Hatzfeld & Williamson 1974, Brunborg & Williamson 1978). This variation from G1 phase to S/G2 has been inferred to be due to differing repair capacities from the absence of relative resistance in repair-deficient strains (Game & Mortimer 1974, Rao & Reddy 1982). The biochemical analysis of the nature of the repair of double-strand DNA breaks has revealed that G1 phase cells are unable to repair such lesions, unlike G2 phase cells (Brunborg *et al.* 1980).

The repair mechanism(s) implicated in the repair of double-strand breaks are recombinational (Resnick 1976) and require a duplicate genome, but as diploid yeast strains still exhibit such cyclic variation and have a duplicate genome then, Brunborg & Williamson (1978) attributed the S/G2 resistance to a sister-chromatid exchange repair mechanism. The sensitivity of cells in G1 phase has been associated with enhanced mutation frequencies (Tippins 1978) which may implicate the lower capacity of cells in G1 phase to repair such ionizing radiation damage by error-free sister-chromatid recombination repair.

Similar fluctuations in sensitivity and mutation induction between G1 and S/G2 phase cells have been found after UV irradiation of synchronized yeast cultures (Esposito 1968, Parry & Cox 1968, Chanet *et al.* 1973, Davies *et al.* 1978). The period of greatest cell sensitivity to the lethal effects of UV was observed in the G1 phase of growth which has been attributed to recominational repair (Fabre 1973), and G1 phase cells exhibited maximum induction of genetic end points. The possibility that the variations observed in reported differences in viability variation for MMS treatment during the mitotic cell cycle to MNNG and other chemical mutagens (Tippins 1979) has not been extended to investigation of genetic effects.

Saccharomyces cerevisiae has proved a useful tool for comparative studies on the sensitivities of mitotic and meiotic cell sensitivities. Differences have been reported such as the mutagenicity of 5-aminoacridine in meiotic cells, in comparision to antimutagenic action in mitotic cells (Magni *et al.* 1964), and the selective toxic effect of the mitochondrial mutagen ethidium bromide to meiotic cells (Kelly & Parry 1984). Variations in cell survival and mutagenesis after UV irradiation and X-ray treatment revealed similar cyclic variations to mitotic cells, with meiotic cells showing increased resistance and reduced mutation frequency during and after meiotic DNA synthesis (Hottinguer-de Margerie & Moustacchi 1979, Kelly & Parry 1983, Kelly *et al.* 1983). For UV treatments other reports have found a UV-sensitive period corresponding to meiotic prophase I (Simchen *et al.* 1973) or no variation (Machida & Nakai 1980); but experiments using the supersporulating strain SK1, and the majority of reports support the conclusion that meiotic cells prior to DNA synthesis and the haploid spore products are most sensitive to UV irradiation (Hottinguer-de Magerie & Moustacchi 1979, Kelly & Parry, 1983, Resnick *et al.* 1983). The sensitivity of the meiotic cells prior to DNA synthesis may not, however, be as great as in mitotic G1 phase cells, but more closely resemble stationary or G0 phase cells which are more resistant (Tippins & Parry 1981).

Chemical mutagenesis in meiosis has also been examined, and again differences have been discovered. No evidence of increased lethal effects of chemical mutagens has been found, but increased levels of mutation in cells prior to meiotic DNA synthesis have been reported for MMS (Kelly & Parry

1986). In the case of MNNG an absence of increased mutagenesis during DNA synthesis in meiosis was also found, although enhanced levels of binding of radiolabel from MNNG was detected in mitochondrial DNA (Kelly *et al.* 1983). Considerable comparative analysis of mutagens, repair characteristics, and mutation mechanisms remains to be performed.

4.11.7 The timing of induced genetic effects (see chapter 3)

The timing of mutagenesis following mutagenic treatments has been the subject of study using pedigree analysis made possible by the micromanipulation of yeast cells (James & Kilbey 1977, James *et al.* 1978, Kilbey & James 1979). These experiments have been performed for UV irradiation with cells in G0 which appear to fix mutations prior to DNA replication in wild-type cells; but in cells without excision repair mutations arise following DNA replication and in subsequent divisions.

Chemical mutagenesis has also been examined in a similar manner (Kilbey 1984). Agents which are not indicated to induce DNA damage which is subject to excision repair, e.g. MNNG and EMS, have a requirement for cell division implicating misreplication mechanisms of mutagenesis. Other mutagen damage which is subject to excision repair, e.g. by 4NQO, MMS, HN_2, DEB, is able to cause mutation in the absence of cell division, albeit at a reduced rate. For these mutations, as after UV irradiation, the fixation of mutations is likely to occur during error-prone excision in which the *RAD6* gene is involved. The timing of fixation of mutation after treatment during other parts of the cell-cycle has not been reported.

4.12 CONCLUSION

Studies on DNA repair and mutagenesis using *Saccharomyces cerevisiae* are at an exciting point of development, as the first phase of molecular cloning studies will enable critical biochemical analysis of the nature of gene products, involved in often rare cellular events. The techniques will be, and are, revealing aspects of regulation, structure, and activity for the yeast genes involved which powerfully complement the classical genetic analysis which has been, and will continue to be, performed.

Studies on excision repair have been described which indicate the interesting model eukaryote system yeast presents by comparison to studies on *Xeroderma pigmentosum* in humans and on the ideal model yeast represents for recombination studies. Regarding mutagenesis, interest lies still in the intriguing nature and role of *RAD6* in mutagenesis. Given the homology of the *RAD6* protein with high-mobility group proteins, which are associated with nucleosome assembly in chromatin, it may be at the level of nucleosome assembly/diassembly that *RAD6* acts (Reynolds *et al.* 1985).

The nature of mutagnesis in yeast has been considered in terms of the central importance of misrepair even for classical misreplication mutagens like MNNG, but mutations occurring by misrepair may occur through modification of repair polymerases or replication fidelity during DNA replication. *RAD6* has been found to be required for the mutagenic effects of excess dTMP (for review see Haynes & Kunz 1982) which supports this model of alteration of replication fidelity during repair or normal DNA replication. Further, the involvement in the error-prone repair pathway of *CDC7* (Njagi & Kilbey 1982) and *CDC8* (Prakash *et al.* 1979) which are affected in DNA synthesis also supports this, although the roles of many of the genes involved may be mediated through subtle side effects. Analysis of genes or gene functions solely involved in mutagenesis offers the prospect of providing clearer information in the future.

REFERENCES

Adzuma, K., Ogawa, T. & Ogdawa, H. (1984). *Mol. Cell. Biol.* **4** 2735–2744.

Akhmedor, A. T., Kaboer, O. K., & Bekker, M. L. (1982). *Biochim. Biophys Acta.* **696** 163–170.

Arlett, C. G. & Lehmann, A. R. (1978). *Ann. Rev. Genet.* **12** 95.

Armel, P. R. & Wallace, S. S. (1978). *Nucl. Acids. Res.* **5** 3347–3356.

Armel, P. R. & Wallace, S. S. (1984). *J. Bacteriol.* **160** 895–902.

Ather, A., Ahmmed, Z., & Riazuddin, S. (1984). *Nucl. Acids. Res.* **12** 2111–2126.

Averbeck, D. & Moustacchi, E. (1975). *Biochim. Biophys. Acta* **395** 393–404.

Averbeck, D., Moustacchi, E., & Bisargni, E. (1978). *Biochim. Biophys. Acta* **518** 464–481.

Bacchetti, S., Casandra, M., & Manro, F. (1967). *Exp. Cell. Res.* **46** 292–300.

Backendorf, C. M., van den Berg, E. A., Brandsma, J. A., Kartasora, T., van Sluis, C. & van de Putte, P. (1983). In: *Cellular responses to DNA damage.* E.C. Friedberg & B. A. Bridges, eds. New York: Alan R. Liss. p. 161.

Bagg, A., Kenyon, C. J., & Walker, G. C. (1981). *Proc. Nat. Acad. Sci. (USA)* **78** 5749–4753.

Baranowska, H. & Putrament, A. (1979). *Mutat. Res.* **63** 291–300.

Barclay, B. J. & Little, J. S. (1978). *Mol. Gen. Genet.* **160** 33–40.

Barker, D. G. & Johnston, L. H. (1983). *Eur. J. Biochem* **134** 315–319.

Barker, D. G., White, J., Johnson, A. L., & Johnston, L. H. (1984). *12th International Conference on yeast genetics and molecular biology,* abstract book, Edinburgh 1984, 57.

Beam, C. A., Mortimer, R. K., Wolfe, R. S., & Tobias, C. A. (1954). *Arch. Biochem. Biophys.* **49** 110–122.

Bennetzen, J. L. & Hall, B. D. (1982). *J. Biol. Chem.* **257** 3026–3031.

Bianchi, L. & Foury, F. (1982). *Mol. Gen. Genet.* **185** 418–423.

Birkenmeyer, L. G., Hill, J. C., & Dumas, L. B. (1984). *Mol. Cell. Biol.* **4** 583–590.

Boatwright, D. T., Madden, J. J., Denson, J., & Werbin, H. (1975). *Biochemistry* **14** 5418–5421.

Brash, D. E. & Haseltine, W. A. (1982). *Nature* **298** 189–192.

Brendel, M. & Haynes, R. H. (1973). *Mol. Gen. Genet.* **125** 197–216.

Brown, A. M. & Kilbey, B. J. (1970). *Mol. Gen. Genet.* **108** 258–265.
Brunborg, G. & Williamson, D. H. (1978). *Mol. Gen. Genet.* **162** 277–286.
Brunborg, S., Resnick, M. A., & Williamson, D. H. (1980). *Radiat. Res.* **82** 547–558.
Burke, W. & Fangman, W. L. (1975). *Cell* **5** 263–269.
Calderon, I. L., Contopoulou, C. R., & Mortimer, R. K. (1982). *Recent Advances in Yeast Molecular Biology* **1** M. S. Espesito (ed.) Noyes Publication: p. 200.
Calderon, I. L., Contopoulou, C. R., & Mortimer, R. (1983). *Curr. Genetics.* **7** 93–100.
Casadaban, M. J., Chou, J., & Cohen, S. N. (1980). *J. Bacteriol.* **143** 971–980.
Cassier, C., Chanet, R., Henniques, J. A. P., & Moustacchi, E. (1980). *Genetics* **96** 841–857.
Cerda-Olmedo, E., Hanawalt, P. C., & Guerda, W. (1968). *J. Mol. Biol.* **33** 705–719.
Chanet, R., Williamson, D. H., & Moustacchi, E. (1973). *Biochim. Biophys. Acta* **324** 290–299.
Chanet, R., Waters, R., & Moustacchi, E. (1975). *Int. J. Radiat. Biol.* **27** 481–845.
Charlesworth, J. D., Chu, Y. H., O'Connor, P. J., & Craig, A. N. (1981). *Carcinogenesis* **2** 329.
Chlebowicz, E. & Jachymczyk, W. K. (1979). *Mol. Gen. Genet.* **167** 179–186.
Chow, T. Y. K. & Resnick, M. A. (1983). In: *Cellular responses to DNA damage,* E. C. Friedberg & B. A. Bridges (eds) New York, Alan R. Liss, 447.
Cox, B. S. & Game, J. C. (1974). *Mutat. Res.* **26** 257–264.
Cox, B. S. & Parry, J. M. (1968). *Mutat. Res.* **6** 37–55.
Culbertson, M. R., Charnas, L., Johnston, M. T., & Fink, G. R. (1977). *Genetics* **86** 745–764.
Davies, P. J., Tippins, R. S., & Parry, J. M. (1978). *Mut. Res.* **51** 327–346.
Dawes, I. W. and Carter, B. L. A. (1974). *Nature* **250** 709–712.
Dicaprio, L. & Cox, B. S. (1981). *Mutat. Res.* **82** 69–85.
Dominski, Z. & Jachymczyk, W. J. (1984a). *Mol. Gen. Genet.* **193** 167–171.
Dominski, Z. & Jachymczyk, W. J. (1984b). *12th International conference on yeast genetics and molecular biology,* abstract book. Edinburgh p. 222.
Donahue, T. F., Farabaugh, P. J., & Fink, G. R. (1980). *Science* **212** 455–457.
Dressler, D. & Potter, H. (1982). *Ann. Rev. Biochem.* **51** 727.
Eckardt, F. & Haynes, R. H. (1977). *Genetics.* **85** 225–247.
Eckardt, F., Kowalski, S., & Laskowski, W. (1975). *Mol. Gen. Genet.* **136** 261–272.
Eckardt, F., Moustacchi, E., & Haynes, R. H. (1978). In: *ICN-UCLA Symposium on Molecular Biology Vol IX DNA Repair Mechanisms,* Hanawat, P. C., Friedberg, E. C. and Fox, C. F. (eds) Acad. Press, New York p. 421.
Ejchart, A. & Putrament, A. (1979). *Mutat. Res.* **60** 173–180.
Esposito, M. A. (1978). *Proc. Nat. Acad. Sci. (USA)* **75** 4436–4440.

Esposito, R. E. (1968). *Genetics* **59** 445–459.
Estereron, A. M. & Sicard, N. (1986). In: *Molecular Biology of DNA repair, Manchester 1986,* poster abstracts.
Fabre, F. (1973). *Radiat. Res.* **56** 528–529.
Fabre, F. (1978). *Nature* **272** 795–798.
Fischer, E., Keijzer, W., Thielmann, H. W., Popanda, O., Bohert, E., Edler, L., Jung, E. G. & Bootsma, D. (1985). *Mutat. Res.* **145** 217–225.
Fogel, S. & Hurst, D. (1963). *Genetics.* **48** 321–328.
Fogliano, M. & Schendel, P. F. (1981). *Nature* **289** 196–198.
Foote, R. S. & Mitra, S. (1984). *Carcinogenesis* **5** 277–282.
Foote, R. S., Mitra, S., & Pal, B. C. (1980). *Biochem. Biophy. Res. Comn.* **97** 654–659.
Forster, J. W. & Strike, P. (1985). *Gene* **35** 71–82.
Foury, F. (1982). *Eur. J. Biochem.* **124** 253–259.
Foury, F. & Goffeau, A. (1979). *Proc. Nat. Acad. Sci. (USA)* **76** 6529–6533.
Friedberg, E. C. (1985a). *DNA repair.* New York: W. H. Freeman & Co.
Friedberg, E. C. (1985b). *Cancer Surveys* Vol. 4. No. 3, 529.
Friedberg, E. C., Fleer, R., Naumorski, L., Nicolet, C. M., Robinson, G. W., Weiss, W. A., & Yang, E. (1986). In: *Mechanisms of antimutagenesis and carcinogenesis.* D. M. Shankel, P. Hartman, T. Kada, & A. Hollaender, eds Plenum Press, New York (in press).
Friedberg, E. C., Ehmann, U. K., & Williams, J. I. (1979). In: *Advances in Radiation Biology,* Lett, J. T. & Adler, H. eds Vol. 8. Academic Press, New York, 85.
Friis, J., Flury, F., & Leupold, U. (1971). *Mutat. Res.* **11** 373–396.
Fujiwara, Y. & Kano, Y. (1983). In: *Cellular responses to DNA damage.* E. C. Friedberg & B. A. Bridges, eds New York, Alan R. Liss, 215.
Game, J. C. (1984). In: *Yeast genetics; fundamental and applied aspects,* Spencer, J. F. T., Spencer, D. M., & Smith, A. R. W. eds Springer-Verlag, p. 109.
Game, J. C. & Cox, B. S. (1971). *Mutat. Res.* **12** 328–331.
Game, J. C. & Cox, B. S. (1972). *Mutat. Res.* **16** 353–362.
Game, J. C. & Cox, B. S. (1973). *Mutat. Res.* **20** 35–44.
Game, J. C. & Mortimer, R. K. (1974). *Mutat. Res.* **24** 281–292.
Gocke, E. & Manney, T. R. (1979). *Genetics* **91** 53–61.
Game, J. C. & Mortimer, R. K. (1974). *Mutat. Res.* **24** 281–292.
Game, J. C. Johnston, L. H., & von Borstel, R. C. (1979). *Proc. Nat. Acad. Sci. USA* **76** 4589–4592.
Golin, J. E. & Esposito, M. S. (1977). *Mol. Gen. Genet.* **150** 127–135.
Golin, J. E. & Esposito, M. S. (1981). *Mol. Gen. Genet.* **183** 252–263.
Gottlieb, D. J. C. & von Borstal, R. C. (1976). *Genetics* **83** 655–666.
Hadden, C. T., Foote, R. S., & Mitra, S. (1983). *J. Bacteriol.* **153** 756–762.
Harris, A., Karran, P., & Lindahl, T. (1983). *Cancer Res.* **43** 3247.
Hastings, P. J., Quah, S., & von Borstal, R. C. (1976). *Nature* **264** 719–722.
Hatzfeld, T. & Williamson, D. H. (1974). *Exp. Cell. Res.* **84** 431–435.
Haynes, R. H. & Kunz, B. A. (1981). *Life cycle and inheritance.* J. Strathern, E. Jones & J. Broach, eds Cold Spring Harbor Lab., New York, 371.
Henriques, J. A. P. & Moustacchi, E. (1980) *Genetics* **95** 273–288.
Higgins, D. R., Prakash, L., Reynolds, P., & Prakash, S. (1984). *Gene* **30** 121–128.

Higgins, D. R., Prakash, S., Reynolds, P., Polakowska, R., Weber, S., & Prakash, L. (1983a). *Proc. Nat. Acad. Sci. (USA).* **80** 5680–5684.

Higgins, D. R., Prakash, S., Reynolds, P., & Prakash, L. (1983b). *Gene* **26** 119–126.

Horice, T. A. & Neale, S. (1984). *Mutat. Res.* **22** 235.

Ho, K. S. Y. (1975). *Mutat. Res* **30** 327–334.

Hunnable, E. G. & Cox, B. S. (1971). *Mutat. Res* **13** 297–309.

Hurst, D. D. & Fogel, S. (1964). *Genetics* **50** 435–458.

Iwatsuki, N., Joe, C. O., & Werbin, H. (1980). *Biochemistry* **19** 1172–1176.

Jackson, J. A. & Fink, G. R. (1981). *Nature* **292** 306–311.

James, A. P. & Kilbey, B. S. (1977) *Genetics* **87** 237–248.

James, A. P., Kilbey, B. S., & Prefontaine, S. J. (1978). *Mol. Gen. Genet.* **165** 207–212.

Johnston, L. H. (1979). *Mol. Gen. Genet.* **170** 327–331.

Johnston, L. H. & Nasmyth, K. A. (1978). *Nature* **274** 891–893.

Johnston, L. H. & Johnston, A. L. (1983). *Mutat. Res* **109** 31–40.

Kacinski, B. M., Sancar, A., & Rupp, W. D. (1981). *Nucl. Acid. Res.* **9** 4495–4508.

Kato, T. & Shinoura, Y. (1977). *Mol. Gen. Genet.* **156** 121–131.

Kee, S. G. & Haber, J. E. (1975). *Proc. Nat. Acad. Sci. (USA)* **72** 1179–1183.

Kelly, S. L. (1982). PhD thesis (Univ. of Wales).

Kelly, S. L. & Parry, J. M. (1983). *Mutat. Res* **108** 109–120.

Kelly, S. L. & Parry, J. M. (1984). *Current Genet.* **8** 69–76.

Kelly, S. L. & Parry, J. M. (1986). *Mutagenesis* **1**(2) 151–155.

Kelly, S. L., Merrill, C., & Parry, J. M. (1983a). *Mol. Gen. Genet.* **191** 314–318.

Kelly, S. L., Tippins, R. S., Parry, J. M., & Waters, R. (1983b). *Carcinogenesis* **4** 851–856.

Kenyon, C. J. & Walker, G. C. (1980). *Proc. Nat. Acad. Sci. (USA)* **77** 2819–2823.

Kenyon, C. J. & Walker, G. C. (1981). *Nature* **289** 808–810.

Kern, R. & Zimmerman, F. K. (1978). *Mol. Gen. Genet.* **161** 81–88.

Khan, N. A., Brendel, M., & Haynes, R. H. (1970). *Mol. Gen. Genet.* **107** 376–378.

Kilbey, B. J. & James, A. P. (1979). *Mutat. Res* **60** 163–171.

Kilbey, B. J., Brychay, T., & Nasim, A. (1978). *Nature* **274** 889–891.

Kilbey, B. J. (1984). *Mol. Gen. Genet.* **197** 519–521.

Kimball, R. F. (1980). *Mutat. Res* **72** 361–372.

Kleihives, P. & Margison, G. P. (1976). *Nature* **259** 153–155.

Krokan, H., Haygen, A., Myrnes, B., & Guddal, P. H. (1983). *Carcinogenesis* **4** 1559.

Kunz, B. A. & Haynes, R. H. (1981). *Ann. Rev. Genet.* **15** 57.

Kunz, B. A. & Glickman, B. (1984). *Genetics* **106** 347–364.

Kuo, C.-L. & Campbell, J. L. (1983). *Mol. Cell. Biol* **3** 1730–1737.

Kupiec, M. & Simchen, G. (1984a) *Curr. Genet.* **8** 559–566.

Kupiec, M. & Simchen, G. (1984b) *Mol. Gen. Genet.* **193** 525–531.

Langeveld, S. A., Yasui, A., & Eker, A. P. M. (1985). *Mol. Gen. Genet.* **199** 396–400.

Lawrence, C. W. (1982). *Adv. Genet.* **21** 173–254.

Lawrence, C. W. & Christensen, R. (1976). *Genetics* **82** 207–232.

Lawrence, C. W. & Christensen, R. (1978a). *J. Mol. Biol* **122** 1–21.
Lawrence, C. W. & Christensen, R. (1978b). *Genetics* **90** 213–226.
Lawrence, C. W. & Christensen, R. (1979). *J. Bacteriol.* **139** 866–876.
Lawrence, C. W. & Christensen, R. (1982). *Mol. Gen. Genet.* **186** 1–9.
Lawrence, C. W. & Christensen, R., & Schwartz, A. (1982). In: *Molecular and Cellular Mechanisms of Mutagenesis. Basic Life Sciences* Lemontt, J. F. & Generoso, W. M. eds, Vol. 20; Plenum, New York; p. 109.
Lawrence, C. W., Nilsson, P. E., & Christensen, R. B. (1985). *Mol. Gen. Genet.* **200** 86–91.
Lawrence, C. W., Stewart, J. W., Sherman, F., & Christensen, R. (1974). *J. Mol. Biol.* **85** 137–162.
Lawrence, C. W., Stewart, J. W., Sherman, F., & Thomas, F. L. (1970). *Genetics* **64** 536–537.
Lemontt. J. F. (1971). *Mutat. Res* **13** 319–326.
Lemontt, J. F. (1972). *Mol. Gen. Genet.* **119** 27–42.
Lemontt, J. F. (1977). *Mutat. Res* **43** 179–204.
Lemontt, J. F. (1980). In: *DNA Repair and Mutagenesis in Eukaryotes.* de Serres, F. J., Generoso, W. M., & Shelby, M. D. eds Plenum Press, New York, p. 85.
Liebman, S. W. & Downs, K. M. (1980). *Mol. Gen. Genet.* **179** 703–705.
Lindahl, T. (1979). *Prog. Nucl. Acid. Res. Mol. Biol.* **22** 135.
Lindahl, T. (1982). *Ann. Rev. Biochem* **51** 61–87.
Lindahl, T., Sedgewick, B., Demple, B., & Karran, P. (1983). In: *Cellular responses to DNA damage.* E.C. Friedberg & B. A. Bridges, eds New York, Alan R. Liss, 241.
Lindamood III C., Bendell, M. A., Billings, K. C., Dyroff, M. C., & Swenberg, J. A. (1983). *Chem. Biol. Interactions* **45** 381.
Lucchini, S., Sora, S., & Panzeri, L. (1980). *Mutat. Res.* **72** 397–404.
Machida, I. & Nakai, S. (1980). *Mutat. Res* **73** 59–68.
MacInnes, M. A., Bingham, J. M., Thompson, L. H., & Striniste, G. (1984). *Mol. Cell. Biol.* **4** 1152.
MacQuillan, M. A., Hermann, A., Coberly, J. S., & Green, G. (1981). *Photochem. Photobiol.* **34** 673.
Maga, J. A. & McEntee, K. (1985). *Mol. Gen. Genet.* **200** 313–321.
Magni, S. E. & Pughii, P. P. (1966). *Cold Spring Harbor Symp. Quant. Biol.* **31** 699–704.
Magni, G. E. & von Borstal, R. C. (1962). *Genetics* **47** 1097–1108.
Magni, G. E., von Borstal, R. C., & Sora, S. (1964). *Mutat. Res* **1** 227–232.
Malone, R. E. & Esposito, R. E. (1980). *Proc. Nat. Acad. Sci. (USA)* **77** 503–507.
Malone, R. E. & Esposito, R. E. (1981). *Mol. Cell. Biol.* **1** 891–901.
Martin, P., Prakash, L., & Prakash, S. (1981). *J. Bacteriol.* **146** 684–691.
Mashell, A. N., Ganges, M. B., Lutzner, M. A., Coon, H. G., Barrett, S. F., Dupuy, J-M., & Robbins, J. H. (1983). In: *Cellular responses to DNA damage.* E. C. Friedberg & B. A. Bridges, eds. A. R. Liss, New York, 209.
Maus, K. L. & Haynes, R. H. (1984). *12th International conference on yeast genetics and molecular biology,* abstract book. 218.
McCarthy, J. G., Edington, B. V., & Schendel, P. F. (1983). *Proc. Natl. Acad. Sci. (USA),* **80** 7380–7384.
McCarthy, T. V., Karran, P., & Lindahl, T. (1984). *EMBO J.* **3** 545–550.

McClanahan, T. & McEntee, K. (1984). *Mol. Cell. Biol.* **4** 235.
McKee, R. H. & Lawrence, C. W. (1979a). *Genetics* **93** 361–373.
McKee, R. H. & Lawrence, C. W. (1979b). *Genetics* **93** 375–381.
McKee, R. H. & Lawrence, C. W. (1980). *Mutat. Res.* **70** 37–48.
Mehta, J. R., Ludlum, D. B. Renard, A., & Verly, W. G. (1981). *Proc. Natl. Acad. Sci. USA,* **78** 6766–6770.
Miller, R. D., Prakash, L., & Prakash, S. (1982a). *Mol. Gen. Genet.* **188** 235–239.
Miller, R. D., Prakash, L., & Prakash, S. (1982b). *Mol. Cell. Biol.* **2** 939.
Montesano, R., Bresil, H., & Margison, G. (1979). *Cancer Res.* **39** 1798–1802.
Montesano, R., Bresil, H., Margison, P., & Pegg, A. E. (1980). *Cancer Res.* **40** 452–458.
Morohashi, F. & Munakata, N. (1983). *Mutat. Res.* **110** 23.
Morrison, D. P. & Hastings, P. J. (1979). *Mol. Gen. Genet.* **175** 57–65.
Mortelmans, K., Friedberg, E. C., Slor, H., Thomas, G., & Cleaver, J. E. (1976). *Proc. Natl. Acad. Sci. USA* **73** 2757–2761.
Moustacchi, E. (1969). *Mutat. Res.* **7** 171–185.
Moustacchi, E. (1971). *Mol. Gen. Genet.* **114** 50–58.
Moustacchi. E. (1972). *Mol. Gen. Genet.* **111** 243–257.
Moustacchi, E., Perlman, P. S., & Mahler, H. R. (1976). *Mol. Gen. Genet.* **148** 251–261.
Mowat, M. & Hastings, P. J. (1979). *Can. J. Genet. Cytol.* **21** 574.
Muhammed, A. (1966). *J. Biol. Chem.* **241** 516–523.
Nagpal, M. L., Higgins, D. R., & Prakash, S. (1985). *Mol. Gen. Genet.* **199** 59–63.
Nakai, S. & Matsumoto, S. (1967). *Mutat. Res.* **4** 129–136.
Naumovski, L., Chu, G., Berg, P., & Freidberg, E. C. (1985). *Mol. Cell. Biol.* **5** 17–26.
Naumovski, L. & Friedberg, E. C. (1982). *J. Bacteriol.* **152** 323–331.
Naumovski, L. & Friedberg, E. C. (1983a). *Proc. Natl. Acad. Sci. (USA).* **80** 4818–4821.
Naumovski, L. & Friedberg, E. C. (1983b). *Gene* **22** 203–209.
Naumovski, L. & Friedberg, E. C. (1984). *Mol. Cell. Biol.* **4** 290–295.
Naumovski, L. & Friedberg, E. C. (1986). *Mol. Cell. Biol.* (in press).
Nicolet, C. M., Chenevert, J. M., & Friedberg, E. C. (1985). *Gene* **36** 225.
Nisson, P. E. & Lawrence, C. W. (1986). *Mutat. Res* **165** 129.
Olsson, M. & Lindahl, T. (1980). *J. Biol. Chem.* **255** 10569–10571.
Orr-Weaver, T. & Szostak, J. (1983). *Proc. Natl. Acad. Sci. (USA).* **80** 4417–4421.
Orr-Weaver, T. & Szostak, J. (1985). *Microbial. Rev.* **49** 35.
Orr-Weaver, T. L., Szostak, J. W., & Rothstein, R. J. (1983). *Methods Enzymol.* **101** 228–245.
Panzeri, L., Sora, S., & Magni, G. E. (1979). *Mutat. Res.* **62** 27–33.
Parry, J. M. & Cox, B. S. (1968). *Mutat. Res.* **5** 373–384.
Parry, J. M., Davies, P. J., & Evans, W. E. (1976). *Mol. Gen. Genet.* **146** 27–35.
Patrick, M. W. & Haynes, R. H. (1968). *J. Bacteriol.* **95** 1350–1354.
Pegg, A. E. (1978). *Biochim. Biophys. Res. Commun.* **84** 166–172.
Pegg, A. E. & Parry, W. (1981). *Carcinogenesis* **2** 1195–1200.

Peterson, T. A., Prakash, L., Prakash, S., Osley, M., & Reed, S. I. (1985). *Mol. Cell. Biol.* **5** 226–235.
Perozzi, G. & Prakash, S. (1986). *Mol. Cell. Biol.* **6** 1497–1507.
Piper, P. W., Wasserstein, M., Engback, F., Kaltoft, K., Celis, J. E., Zeuthen, J., Liebman, S., & Sherman, F. (1976). *Nature* **262** 757–761.
Polakowska, R. & Putrament, A. (1979). *Mutat. Res.* **61** 207–213.
Prakash, L. (1974). *Genetics* **78** 1101–1118.
Prakash, L. (1975a). *J. Mol. Biol.* **98** 781–795.
Prakash, L. (1976a). *Genetics* **83** 285–301.
Prakash, L. (1976b). *Mutat. Res.* **41** 241–248.
Prakash, L. (1977a). *Mutat. Res.* **45** 13–20.
Prakash, L. (1977b). *Mol. Gen. Genet.* **152** 125–128.
Prakash, L. (1981). *Mol. Gen. Genet.* **184** 471–478.
Prakash, L. & Prakash, S. (1977). *Genetics* **86** 33–55.
Prakash, L. & Prakash, S. (1979). *Mol. Gen. Genet.* **176** 351–359.
Prakash, L. & Sherman, F. (1973). *J. Mol. Biol.* **79** 65–82.
Prakash, L. & Sherman, F. (1974). *Genetics* **77** 245–254.
Prakash, L. & Taillon-Miller, R. (1981). *Current Genet.* **3** 247–250.
Prakash, L. & Higgins, D. (1982). *Carcinogenesis,* **3** 439.
Prakash, L., Stewart, J. W., & Sherman, F. (1974). *J. Mol. Biol.* **79** 65–82.
Prakash, L., Hinkle, D., & Prakash, S. (1979). *Mol. Gen. Genet.* **172** 249–258.
Prakash, L., Prakash, L., Burke, W., & Montelone, B. A. (1980). *Genetics* **94** 31–50.
Prakash, L., Polakawska, R., & Slitzky, B. (1982). In: *Recent Advances in Yeast Molecular Biology* 1, Esposito, M. J. ed. Noyes Publications p. 225.
Quah, S., von Borstel, R. C., & Hastings, P. J. (1980). *Genetics* **96** 819–839.
Rao, B. S. & Reddy, N. M. (1982). *Mutat. Res.* **95** 213–224.
Resnick, M. A. (1969). *Genetics* **62** 519–531.
Resnick, M. A. (1975). In: *Molecular mechanisms for repair of DNA*. P.C. Hanawatt & R. M. Setlow, eds. Part B, Plenum Press, New York: 549.
Resnick, M. A. & Setlow, J. K. (1972). *J. Bacteriol.* **109** 979–986.
Resnick, M. A. & Martin, P. (1976). *Mol. Gen. Genet.* **143** 119–129.
Resnick, M. A., Game, J. C., & Stasiewicz, S. (1983a). *Genetics* **104** 603–618.
Resnick, M. A., Stasiewicz, S., & Game, J. C. (1983b). *Genetics* **104** 583–601.
Reynolds, P., Higgins, D. R., Prakash, L., & Prakash, S. (1985a). *Nucl. Acid. Res.* **13** 2357–2372.
Reynolds, P., Weber, S., & Prakash, L. (1985b). *Proc. Natl. Acad. Sci. (USA)*. **82** 168–172.
Reynolds, R. J. (1978). *Mutat. Res.* **50** 43–56.
Reynolds, R. J. & Friedberg, E. C. (1980). In: *DNA repair and mutagenesis in eukaryotes,* W. M. Generoso, M. D. Shelby, & F. J. Desenes eds. Plenum Press, New York. 121.
Reynolds, R. J. & Friedberg, E. C. (1981). *J. Bacteriol.* **146** 692–704.
Reynolds, R. J., Love, J. D., & Friedberg, E. C. (1981). *J. Bacteriol.* **147** 705–708.
Robins, P. & Cairns, J. (1979). *Nature* **280** 74–76.

Robinson, G. W., Nicolet, C. M., Kalainor, D., & Friedberg, E. C. (1986). *Proc. Natl. Acad. Sci. USA* **83** 1842–1846.
Roman, H. & Fabre, F. (1983). *Proc. Nat. Acad. Sci. USA* **80** 6912–6916.
Rothstein, R. (1979). *Cell.* **17** 185–190.
Rubin, J. S. & Joyner, A. L., *Nature* **306** 206–208.
Ruby, S. W. & Szostak, J. W. (1985). *Mol. Cell. Biol.* **5** 75–84.
Ruhland, A. & Brendel, M. (1979). *Genetics* **92** 83–97.
Ruhland, A., Hasse, E., Siede, W., & Brendel, M. (1981). *Mol. Gen. Genet.* **181** 346–354.
Saeki, T., Machida, I., & Nakai, S. (1980). *Mutat. Res* **73** 251–265.
Samson, L. & Cairns, J. (1977). *Nature* **267** 281–283.
Schild, D., Calderon, I. L., Contopoulou, R., & Mortimer, R. K. (1983a). In: *Cellular responses to DNA damage.* E. C. Friedberg & B. A. Bridges, eds. New York, Alan R. Liss, 417.
Schild, D., Konforti, B., Perez, C., Gish, W., & Mortimer, R. K. (1982). In: *Recent Advances in Yeast Molecular Biology,* Noyes Publication ed. Esposito, M. S. 213.
Schild, D., Konforti, B., Perez, C., Gish, W., & Mortimer, R. K. (1983). *Current Genetics* **7** 85–92.
Schild, D., Johnston, J., Chang, C., & Mortimer, R. K. (1984). *Mol. Cell. Biol.* **4** 1864.
Schild, D., Konforti, B., Perez, C., Gish, W., & Mortimer, R. K. (1982). In: *Recent Advances in yeast Molecular Biology* **1** Nyes Publication, ed. Esposito M. S.: 213.
Schwartz, D. C. & Cantor, C. R. (1984). *Cell.* **37** 67–75.
Schwencke, J. & Moustacchi, E. (1982a). *Mol. Gen. Genet.* **185** 290–295.
Schwencke, J. and Moustacchi, E. (1982b). *Mol. Gen. Genet.* **185** 296–301.
Sclafani, R. A. & Fangman, W. L. (1984). TK. *Proc. Natl. Acad. Sci. USA* **81** 5821–5825.
Sedgwick, B. & Lindahl, T. (1982). *J. Mol. Biol.* **154** 169–175.
Setlow, J. K. & Boiling, M. E. (1963). *Photochem. Photobiol.* **2** 471.
Sherman, F., Stewart, J. W., Jackson, H., Gilmore, R. A., & Parker, J. H. (1974). *Genetics* **77** 255–284.
Shine, J. & Delgarno, L. (1974). *Proc. Nat. Acad. Sci. (USA).* **71** 1342–1346.
Siede, W. & Eckardt, F. (1984). *Mutat. Res.* **129** 3–11.
Siede, W., Eckardt, F., & Brendel, M. (1983a). *Mol. Gen. Genet.* **190** 406–412.
Siede, W., Eckardt, F., & Brendel, M. (1983b). *Mol. Gen. Genet.* **190** 413–416.
Simchen, G., Salts, Y., & Pinnon, R. (1973). *Genetics* **73** 531–541.
Siminovitch, L. (1976). *Cell.* **7** 1–11.
Sklar, R. & Strauss, B. (1980). *J. Mol. Biol.* **143** 343–362.
Slominski, P. P., Perrodin, G., & Croft, H. H. (1968). *Biochem. Biophys. Res. Commun.* **30** 232–239.
Snow, R. (1967). *J. Bacteriol.* **94** 571–575.
Stern, H., & Holta, Y. (1973). *Ann. Rev. Genet.* **7** 37–66.
Szostak, J. W., Orr-Weaver, T. L., Rothstein, R. J., & Stahl, F. W. (1983). *Cell.* **33** 25–35.
Thompson, L. H., Brookman, K. W., Dillehay, L. E., Mooney, C. L., & Carrano, A. V. (1982). *Somatic Cell. Genet.* **8** 759.

Thompson, L. H., Brookman, K. W., Minkler, J. L., Fuscoe, J. C., Henning, K. H., & Carrano, A. V. (1985a). *Mol. Cell. Biol.* **5** 881–884.
Thompson, L. H., Mooney, C. L., & Brookman, K. W. (1985b). *Mutat. Res.* **150** 423–429.
Tippins, R. S. (1979). PhD. thesis (University of Wales).
Tippins, R. S. & Parry, J. M. (1981). *Int. J. Radiat. Biol.* **40** 327–331.
Topal, M. D., Hutchinson, C. A., & Baker, M. S. (1982). *Nature* **298** 963–965.
Tuite, M. F. & Cox, B. S. (1981). *Mol. Cell. Biol.* **1** 153–157.
Unrau, P., Wheatcroft, R., & Cox, B. S. (1971). *Mol. Gen. Genet.* **113** 359–362.
Van Duin, M., de Wit, J., Odijk, H., Westerveld, A., Yasui, A., Koken, M., Hoejmakers, J. H. J., & Bootsma, D.)1986). *Cell.* (in press).
Warren, W. & Lawley, P. D. (1980). *Carcinogensis,* **1** 67.
Waters, R. & Moustacchi, E. (1974a). *Biochim. Biophys. Acta* **353** 407–419.
Waters, R. & Moustacchi, E. (1974b). *Biochim. Biophys. Acta* **366** 241–250.
Weiss, W. A. & Friedberg, E. C. (1985). *EMBO J.* **4** 1575.
Werbin, H. & Madden, J. J. (1975). *Biochim. Biophys. Acta* **383** 160–167.
Werbin, H. & Madden, J. J. (1977). *Photochem. Photobiol.* **25** 421.
Westerveld, A., Hoejmakers, J. H. H., van Duin, M., de Wit, J., Odijk, H., Pastrick, A., Wood, R. D., & Bootsma, D. (1984). *Nature* **310** 425–428.
Whelan, W. L., Gocke, E., & Manney, T. R. (1979). *Genetics* **91** 35–51.
White, C. I. & Sedgwick, S. G. (1985). *Mol. Gen. Genet.* **201** 99.
Wilcox, D. R. & Prakash, L. (1981). *J. Bacteriol.* **148** 618–623.
Witkin, E. M. (1974). *Proc. Nat. Acad. Sci. (USA)* **71** 1930–1934.
Yang, E. & Friedberg, E. C. (1984). *Mol. Cell. Biol.* **4** 2161.
Yasui, A. & Chevallier, M. R. (1983). *Curr. Genet.* **7** 191.
Yasui, A., Langereld, S. A., & Eker, A. P. M. (1984). *12th International conference on yeast genetics and molecular biology, abstract book,* Edinburgh, 212.
Zimmermann, F. K. (1968). *Mol. Gen. Genet.* **102** 247–256.

5

Yeast cytochrome P-448 enzymes and the activation of mutagens, including carcinogens

Dr. David J. King† and Dr. Alan Wiseman
Biochemistry Division, Department of Biochemistry, University of Surrey, Guildford

5.1 CYTOCHROME P-448/P-450 ENZYMES

5.1.1 Introduction

Many toxic and carcinogenic chemicals are active only after undergoing metabolism to highly electrophilic reactive intermediates. Many of the carcinogens and mutagens which require such activation are metabolized by microsomal monooxygenases of the cytochrome P-450/P448 type. Cytochrome P-450 enzymes are a family of haemoproteins with spectral peaks close to 450 nm (reduced iron: carbon monoxide complex), which include the cytochrome P-448 enzymes and which catalyse a wide range of monooxygenase reactions. These reactions include the monooxygenation of many xenobiotic compounds such as drugs, carcinogens and mutagens (putative carcinogens), and also endogenous compounds such as fatty acids and steroids. Some cytochrome P-450 enzymes have a wide substrate specificity whilst others are dedicated to one or a narrow range of substrates.

Cytochrome P-450 enzymes are widely distributed throughout nature and have been demonstated in both prokaryotic and eukaryotic microorganisms. In microbial systems cytochrome P-450 enzymes appear to have narrower substrate specificities.

The reaction catalysed by cytochrome P-450 can be represented in simple form as:

$$SH + NADPH + H^+ + O_2 \rightarrow SOH + NADP^+ + H_2O$$

where SH represents substrate and SOH the monooxygenated product. The nature of the reaction obviously depends on the chemical nature of the

---

†Present address: Department of Protein Biochemistry, Celltech Ltd., 244–250 Bath Road, Slough, Berks, UK SL14DY.

substrate but can be, for example, aromatic ring hydroxylation, hydroxylation of an aliphatic side chain, dealkylation of a secondary or tertiary amine, deamination, dehalogenation, or C-demethylation. In all of these cases the first step can be visualized as an hydroxylation (Gillette 1966). More than one enzyme is usually involved in the metabolism of a xenobiotic. For example, in the aromatic hydroxylation of benzo(a)pyrene, the action of cytochrome P-450 results in the formation of epoxides which are converted into hydroxylated products by the action of epoxide hydrase (Sims *et al.* 1974).

5.1.2 Mammalian cytochrome P-450 enzymes and their reaction mechanism

Within mammalian tissues cytochrome P-450 enzymes are most abundant in the liver, but they have also been demonstrated in most other body tissues. Most of the cytochrome P-450 of a eukaryotic cell is located in the endoplasmic reticulum which is usually isolated in the form of small vesicles termed microsomes. There are also cytochrome P-450 systems in mammalian mitochondria, particularly in the liver and adrenal cortex. These mitochondrial cytochrome P-450 enzymes are involved in steroid metabolism, and although membrane bound they operate *via* a similar electron transport mechanism to that found in bacterial soluble cytochrome P-450 systems. Cytochrome P-450 enzymes have also been demonstrated in the nuclear envelope, which may be important in the metabolism of carcinogens (Bresnick 1978) and also in the cell membrane (Stasiecki *et al.* 1980).

The components of the microsomal monooxygenase system are well established. Two proteins are involved, a cytochrome P-450 (a haemoprotein containing the protoporphyrin IX group) linked to NADPH: cytochrome P-450 reductase, a flavoprotein containing one FAD and one FMN group. The cytochrome P-450 component contains the substrate and oxygen binding sites, whilst the reductase serves to transport electrons from NADPH to the cytochrome P-450 component, forming an electron transport chain as shown in Fig. 5.1. Phospholipid is also required for a fully

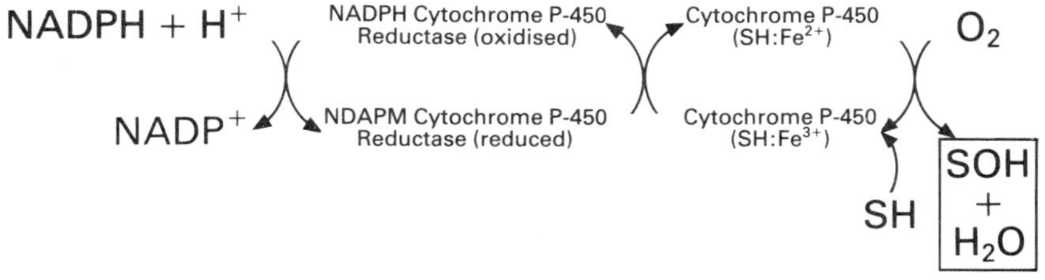

NADPH + H^+
NADPH Cytochrome P-450 Reductase (oxidised)
Cytochrome P-450 (SH:Fe^{2+})
O_2
$NADP^+$
NDAPM Cytochrome P-450 Reductase (reduced)
Cytochrome P-450 (SH:Fe^{3+})
SOH + H_2O
SH

Fig. 5.1 — Electron transport chain for microsomal cytochrome P-450. SH = substrate, SOH = hydroxylated product.

functional system. The role of this phospholipid is unknown although it may have a structural role in associating the protein components together. These components have been purified and reconstituted into an active system by many workers. This was first achieved by Lu & Coon (1968) who reconstituted a system capable of oxidizing fatty acids, steroids, drugs, and other xenobiotics from components isolated from rabbit liver microsomes. The phospholipid requirement can be replaced in many systems by non-ionic detergents, suggesting a relatively non-specific role for the phospholipid.

Although only three components are necessary for the reconstitution of a functional monooxygenase system, the situation *in vivo* is usually more complex. In addition to the NADPH-dependent cytochrome P-450 system, liver microsomes also contain an NADH-dependent electron transport system (involved in fatty acid desaturation) consisting of a haemoprotein, cytochrome b_5 and a flavoprotein, NADH: cytochrome b_5 reductase (Shimakata *et al.* 1972). Lipid is also essential for this system (Holloway 1971). NADH alone cannot support the microsomal monooxygenation of drugs, but when both NADPH and NADH are present a synergistic effect is usually observed, with a faster rate of monooxygenation being achieved than with NADPH alone.

Cytochrome b_5 has been shown to be required for maximal activity of several, though not all, cytochrome P-450 reactions including chlorobenzene hydroxylase (Lu *et al.* 1974), testosterone 16α-hydroxylase (Lu *et al.* 1974), 7-ethoxycoumarin deethylase (Imai 1979), benzo(a)pyrene hydroxylase (Brunstrom & Ingelman-Sundberg 1980), and *p*-nitroanisole-O-demethylase (Sugiyama *et al.* 1980). The synergistic effect of NADH in these cases is thought to be due to the cytochrome b_5-mediated donation of the required 'second electron' to cytochrome P-450 during the catalytic cycle (Hildebrandt & Estabrook 1971, Correira & Mannering 1973). However, cytochrome b_5 is not an absolute requirement for activity in most cases (although required for maximum rate), so the second electron may also come from NADPH via NADPH: cytochrome P-450 reductase when cytochrome b_5 is absent. This is shown clearly in the NADPH-supported purified reconstituted systems mentioned above.

Not all cytochrome P-450 activities are stimulated by the cytochrome b_5 system, and in some cases the activity is reduced owing to competition for electrons (Bosterling *et al.* 1982). It seems, therefore, that for some cytochrome P-450-dependent monooxygenations, cytochrome b_5 is involved in transfer of the second electron, yet for other activities this is not the case. Immunological evidence also supports a role for cytochrome b_5 in only some cytochrome P-450-mediated reactions (Noshiro *et al.* 1980). The occurrence of multiple forms of cytochrome P-450 is now very well established (see on), and it seems that some forms of cytochrome P-450 use cytochrome b_5-mediated transfer of the second electron for oxygenation of some substrates, whereas others use NADPH: cytochrome P-450 reductase for the donation of both electrons (Chiang 1981, Bosterling *et al.* 1982).

The mechanism of the cytochrome P-450 monooxygenase reaction is not fully understood, but the major sequence of events is known, as is shown in

Fig. 5.2. The mechanism involves firstly binding of substrate to the ferric iron form of the enzyme, followed by one electron transfer from NADPH via NADPH: cytochrome P-450 reductase. Oxygen is then bound to this complex and another one-electron transfer occurs from either NADPH via NADPH: cytochrome P-450 reductase or NADH via NADH: cytochrome b_5 reductase and cytochrome b_5. This leads to the activation of oxygen so that reaction with the substrate can occur. The nature of the active oxygen species is currently the subject of extensive research but is thought to involve a number of radical species (White & Coon 1980). The product then dissociates from the complex regenerating the ferric haemoprotein.

The mechanism of cytochrome P-450 reactions would be expected to produce a stoichiometry of one mole of NADPH, and one mole of oxygen consumed per mole of product formed. However, with both microsomal and purified reconstituted systems this is often not observed, and excessive amounts of oxygen and NADPH are consumed (Powis & Jansson 1979). This has been shown to be associated with concomitant production of peroxide and it is now believed that this effect is due to 'uncoupled' cytochrome P-450 activity which uses oxygen and NADPH to form hydrogen peroxide (Lichtenberger & Ullrich, 1977). Thus the expected stoichiometry of the monooxygenase reaction is not seen, owing to the extra oxygen and electrons used up in this peroxide-producing pathway.

Some forms of cytochrome P-450 may exert their monooxygenase reactions through free active oxygen species generated by the enzyme in the absence of bound substrate. This is thought to be the case for the ethanol-inducible cytochrome P-450 in mammalian liver (Ingelman-Sundberg & Hagbjork, 1982).

The requirement of the cytochrome P-450-dependent monooxygenase system for NADPH and oxygen can be replaced by a variety of artificial oxygen and electron donors. These include a number of peroxides. Hrycay & O'Brien (1972) were the first to demonstrate this cytochrome P-450 shunt mechanism with peroxides, by measuring rates of tetramethyl-*p*-phenylenediamine oxidation with cytochrome P-450 and various organic hydroperoxides. Kadlabur *et al.* (1973) demonstrated that cytochrome P-450 could catalyse the oxidation of several amine substrates when NADPH was replaced by a variety of organic hydroperoxides. The highest rate of oxidation was observed with cumene hydroperoxide. Rahimtula & O'Brien (1974) showed that cytochrome P-450-dependent hydroxylation of aniline, benzo(a)pyrene, coumarin, and biphenyl could be supported by cumene hydroperoxide in rabbit liver microsomes. Cumene hydroperoxide will also support cytochrome P-450-dependent fatty acid hydroxylation in liver microsomes (Ellin & Orrenius 1975).

Other oxidising agents also, have been described which can act as artificial oxygen and electron donors to cytochrome P-450 in the same way as peroxides. These include sodium periodate and to a lesser extent sodium chlorite (Gustafsson *et al.* 1979). Nordblom *et al.* (1976) demonstrated that a purified rabbit liver microsomal cytochrome P-450 could metabolize benzphetamine and other substrates in the absence of NADPH and reductase

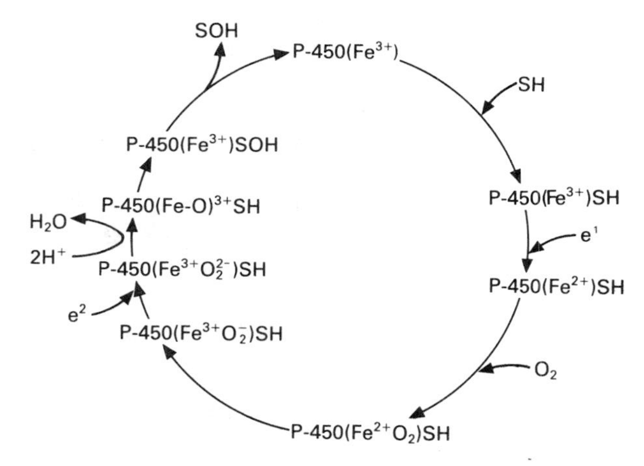

Fig. 5.2 — The cytochrome P-450 reaction cycle. e^1 is transferred form NADPH *via* NADPH: cytochrome P-450 reductase, e^2 is transferred either as is e^1 *or* from NADPH *via* NADPH: cytochrome b_5 reductase and cytochrome b_5.

when certain peroxides were added to the system, although the lipid requirement was retained. In this system several alkyl hydroperoxides, hydrogen peroxide, peroxy acids, and sodium chlorite were all shown to be capable of supporting benzphetamine demethylation by the purified cytochrome P-450, although the high concentrations of some of these compounds necessary to achieve benzphetamine demethylation also resulted in the irreversible oxidation of the cytochrome P-450 haem group, leading to destruction of the enzyme after a few minutes. Rahimtula *et al.* (1978) compared cytochrome P-450 reactions supported by NADPH with cumene hydroperoxide-supported reactions, and found that a common reaction mechanism occurred in both cases. Capdevila *et al.* (1980) also compared the reaction mechanism of cytochrome P-450-dependent monooxygenation when supported by NADPH and cumene hydroperoxide. These workers demonstrated that different patterns of benzo(a)pyrene metabolites were produced in the NADPH and cumene hydroperoxide-supported reactions, with a preponderance of phenols and dihydrodiols in the former and a preponderance of quinones with the latter. These results raise doubts as to the essential similarity of the NADPH and cumene hydroperoxide-supported reactions, although of course it is difficult to rule out non-specific attack by peroxide species. MacDonald *et al.* (1982) have presented evidence that the cytochrome P-450-catalysed hydroxylation of cyclohexane is achieved by a similar mechanism when supported by NADPH or iodosobenzene.

Several groups have studied the interaction between the components of the monooxygenase system and their orientation in the microsomal membrane. The microsomal membrane, which is formed by a 'pinching off' process from the endoplasmic reticulum during homogenization, is believed to retain the same orientation as the endoplasmic reticulum: with evidence for this coming from the localization of several marker enzymes (De Pierre & Ernster 1980). More than three-quarters of the NADPH: cytochrome P-450 reductase molecule is believed to be sited off the lipid bilayer. Cytochrome P-450 molecules, however, are deeply embedded in the membrane, thus making solubilization of these haemoproteins very difficult (Vermillion & Coon 1978). Consistent with these results, De Pierre & Ernster (1980) found that cytochrome P-450 is associated with both sides of the microsomal membrane, and claimed therefore that cytochrome P-450 traverses the lipid membrane whereas NADPH: cytochrome P-450 reductase is only associated with the inside (cytoplasmic) surface. Cytochrome b_5 and NADH: cytochrome b_5 reductase are also associated with the cytoplasmic surface, and are also thought to be largely free of the membrane with only a small part of the molecule embedded in it (De Pierre & Ernster 1980).

The stoichiometry (ratio) of cytochrome P-450 molecules to NADPH:-cytochrome P-450 reductase molecules in the intact endoplasmic reticulum is between 10:1 and 100:1 (Estabrook *et al.* 1971, Sato & Omura 1978). Nebert (1979) has suggested that one cytochrome P-450 molecule may donate electrons to another form of cytochrome P-450 as in mitochondrial electron transport. Two models of the membrane arrangement of cytochrome P-450 and its reductase have been proposed: a rigid model where 8–12 molecules of cytochrome P-450 are arranged in a cluster around the reductase molecule (Peterson *et al.* 1976), and a non-rigid model allowing free diffusion of both enzymes in the lateral plane of the membrane (Yang 1975). There is some experimental data for both models, yet there is little reliable evidence to choose between them.

When cytochrome P-450 is purified, the enzyme exists as a hexamer in solution. When it is reconstituted with the reductase it forms a molecular aggregate with an apparent molecular weight of approx. 800 000, the two proteins combining in a 1:1 molar ratio (French *et al.* 1980). McIntosh *et al.* (1980) concluded from crosslinking and rotational diffusion studies that cytochrome P-450 has a tendency to form aggregates in the microsomal membrane. Schwarz *et al.* (1982) have presented evidence for the cluster-like organization of cytochrome P-450 molecules around the reductase in the microsomal membrane, from studies with saturation transfer EPR spectroscopy showing a very slow rotation of cytochrome P-450 molecules in the form of large aggregates.

The binding of substrates and other compounds to cytochrome P-450 has been widely studied, using the technique of difference spectrophotometry (Schenkman *et al.* 1967). Three different types of spectrum result from binding interactions with cytochrome P-450: type I (spectral maximum 385–390 nm, minimum 420 nm), type II (spectral maximum 425–430 nm, minimum 390–410 nm), and reverse type I which is the mirror-image of the type I spectrum. These spectral changes on binding are not fully understood. The type I spectrum is thought to result from binding of a substrate to the substrate binding site in the cytochrome P-450 molecule. The type II spectrum is due to the binding of a compound to a site near the haem group, possibly at the site of the fifth ligand to the haem iron, occupied by oxygen during catalysis (Schenkman *et al.* 1981). The causes of the spectral changes are now interpreted in terms of spin–state changes and are discussed in section 5.3.

The active cytochrome P-450 molecule contains a haem group with iron in the ferric form. This means that the iron atom has five electrons in the outer d-orbital and, depending on the extent of spin-pairing of these, can exist in two spin states, high spin and low spin. The low spin form results when four of the five d-electrons are paired and corresponds to a six-coordinated haem iron. The high spin form results when the five electrons are in separate energy levels and not paired, and corresponds to a five-coordinated haem iron. In intact microsomal membranes a mixture of the two spin states occurs with approximately equal amounts of each, the ratio of the two forms being in a temperature dependent equilibrium (Cinti *et al.* 1979).

It has been shown that type I substrates of cytochrome P-450 cause a change in the spin state *in vivo* from low to high spin (Kumaki *et al.* 1978). Ristau *et al.* (1979) also, have demonstrated that substrates induce a spin state transition from low to high spin. It has also been shown that the spin state of cytochrome P-450 controls the redox potential of the molecule, with the high spin form having a less negative redox potential (Sligar 1976). Substrate binding thus not only results in a transition from low to high spin but also a change in the redox potential, making it less negative and hence allowing electrons to flow to the cytochrome P-450 molecule more easily (Sligar *et al.* 1979). Thus the substrate-induced spin state change can result in an acceleration of the rate of reduction of the Fe^{3+} of the cytochrome (Misselwitz *et al.* 1980). This reveals a mechanism whereby the substrate facilitates the electron flow to cytochrome P-450, hence enabling the reaction to proceed. It is now known that cytochrome b_5 and NADPH: cytochrome P-450 reductase can also modulate the cytochrome P-450 spin state (Tamburini & Gibson 1983) as can membrane phospholipids (Gibson *et al.* 1980), and all of these factors may exert a subtle control mechanism on the cytochrome P-450 enzyme (Gibson 1985).

5.1.3 The induction of mammalian cytochrome P-450 enzymes and the presence of multiple forms

The activity of the liver microsomal monooxygenase system can be altered by the treatment of the animal with a wide variety of compounds including drugs and carcinogens. This change in activity results from an alteration in the relative amounts of cytochrome P-450 isozymes produced by the animal, usually due to the induction of one or more forms of the enzyme. Induction occurs by *de novo* synthesis of the induced enzymes, resulting from the induction of specific messenger RNA species (Kumar & Padmanaban 1980). This change in the relative amounts of cytochrome P-450 isozymes is reflected in a change in the pattern of metabolism of xenobiotic compounds.

Two of the most widely studied inducers of mammalian liver cytochrome

P-450 enzymes are phenobarbital and 3-methylcholanthrene. After induction with these compounds, different forms of cytochrome P-450 are observed, each with different substrate specificities. The major isozyme present after phenobarbital pretreatment has a very wide substrate specificity and results in the metabolism and detoxication of many xenobiotic compounds. The major isozyme after pretreatment with 3-methylcholanthrene and other compounds such as benzo(a)pyrene and 2,3,7,8-tetrachlorodibenzo-*p*-dioxin (TCDD) is an enzyme with relatively narrow specificity with a Soret peak in the reduced carbon monoxide difference spectrum at 448 nm. Forms of cytochrome P-450 of this type are often known as cytochrome P-448 enzymes, and are very potent for the activation of carcinogens (see section 5.1.4). Several hundred inducers of cytochrome P-450 enzymes are now known, some of which are 'phenobarbital-like' or '3-methylcholanthrene-like', and others such as safrole and pregnenolone 16α-carbonitrile which induce different ranges of isoenzymes. The exact number of cytochrome P-450 isoenzymes present in mammalian liver is unknown, although current estimates range up to about 20 isoenzymes (Levin *et al.*1985, Coon *et al.* 1985). Many of these isoenzymes have now been purified and characterized chemically and immunologically (for example, Levin *et al.* 1985, Ozols 1986, Sesardic *et al.* 1986). As yet no uniform nomenclature has been adopted, and thus several different names may exist for one cytochrome P-450 isoenzyme, leading to some confusion in the literature.

It is now thought that the range of cytochrome P-450 isoenzymes present in mammalian systems includes some with wide substrate specificities such as those in mammalian liver induced by phenobarbital, which are capable of the metabolism of a wide range of xenobiotics, and also some which are dedicated to one or a narrow range of substrates. Examples of dedicated cytochrome P-450 enzymes include the well-characterized mitochondrial enzyme cytochrome P-450_{scc} which is responsible for the side-chain cleavage of cholesterol (Takemori *et al.* 1978), the microsomal cytochrome P-450_{c21} which is specific for steroid 21-hydroxylation (White *et al.* 1984), and thromboxane synthetase (Hauraud & Ullrich 1985).

Recently a great deal of work on the characterization of different cytochrome P-450 isoenzymes has been carried out, including the sequencing of several structural genes (Fujii-Kuriyama *et al.* 1982, Sogawa *et al.* 1984), and amino acid sequencing of several of the cytochrome P-450 proteins (Haniu *et al.* 1984, Ozols 1986). These studies have shown that the cytochrome P-450 isoenzymes can be grouped into families, where one or more members of each family are present in each species (Fujii-Kuriyama *et al.* 1984). There is considerable structural homology between members of the same family, and much less between families which often differ considerably (Fujii-Kuriyama *et al.* 1984). There is now considerable research effort into attempts to establish relationships between different cytochrome P-450 isoenzymes both within and between these enzyme families (Sesardic *et al.* 1986; Phillips *et al.* 1985, Nebert & Gonzalez 1985, Yoshioka *et al.* 1986).

The effect of inducers is now thought to be in inducing a range of cytochrome P-450 isoenzymes, the combined properties of which make up the characteristic properties of the 'induced enzyme'. At the same time the synthesis of other cytochrome P-450 enzymes is reduced. Some of the induced enzymes are present at lower levels in the uninduced animal. The change in the levels of induced or repressed cytochrome P-450 enzymes is due to changes in the mRNA level, as shown by Shephard *et al.* (1982). These workers showed that administration to rats of phenobarbital, resulted in an increase in the mRNA level for PB-P-450. When rats were administered with β-naphthoflavone the level of mRNA for PB-P-450 was reduced.

The major cytochrome P-450 induced in mouse liver by polycyclic aromatic hydrocarbons such as 3-methylcholanthrene, benzo(a)pyrene, or TCDD is a cytochrome P-448, and is often termed cytochrome P_1-450 (Nebert & Gonzalez 1985). The equivalent cytochrome P-450 species in rat liver is often termed cytochrome P-450_c (Ryan *et al.* 1982), with the corresponding enzyme from rabbit liver known as cytochrome P-450 LM_4 based on its electrophoretic mobility (Coon *et al.* 1980).

The induction of mouse cytochrome P_1-450 has been extensively studied by Nebert and co-workers and has been shown to be co-induced with another isoenzyme, cytochrome P_3-450 (Nebert & Gonzalez 1985). Induction is mediated through the 'Ah locus' which is a combination of regulatory, structural, and possibly temporal genes. The hydrophobic inducing compound enters the cell and binds with high affinity to a cytosolic receptor, the Ah receptor. The inducer–receptor complex then undergoes temperature-dependent translocation into the nucleus, and activates transcription of the P_1-450 and P_3-450 genes leading to increases in the enzyme levels in the endoplasmic reticulum (Gonzalez *et al.* 1984, Nebert *et al.* 1984). The mechanism of this transcriptional activation is now under intensive study and is very complex. Gonzalez & Nebert (1985) have identified several functional regions in the upstream region of the mouse cytochrome P_1-450: a promoter region; a region that is negatively autoregulated; possible repressor-binding and inducer–receptor complex binding sites; and an upstream activation element that is required for transcriptional activation by TCDD.

A similar mechanism has been proposed for the equivalent enzyme in rat liver, cytochrome P-450_c (Iversen *et al.* 1986). In this case two binding proteins have been identified which appear to be involved in the induction process (Tierney *et al.* 1983). The expression of the rat cytochrome P-450_c gene is turned on rapidly, with the primary transcript detectable within 15 minutes (Foldes *et al.* 1985), and the cytochrome P-450_c protein is increased steadily for the first 1–2 days after exposure to the inducer (Harada & Omura 1983). There are now known to be at least five distinct cytochrome P-450 gene families (Nebert & Gonzalez 1984), and the induction mechanism outlined above has been demonstrated for only the polycyclic aromatic hydrocarbon inducible cytochrome P-450 enzymes. It is quite possible that different induction mechanisms may exist for other types of cytochrome P-450 enzymes, although little is known about these at present (Burnet *et al.* 1986).

5.1.4 Cytochrome P-450 enzymes in the activation of carcinogens

Many carcinogens such as polycyclic aromatic hydrocarbons have been shown to express their carcinogenicity only after metabolism, by cytochrome P-450 enzymes, to reactive intermediates (Nelson 1982). As mentioned above, cytochrome P-450 enzymes of the cytochrome P-448 type, such as the mouse cytochrome P_1-450 or rat cytochrome P-450_c, have been shown by many workers to be those cytochrome P-450 isoenzymes responsible for the activation of carcinogens. For example, using the Ames test (bacterial system) with benzo(a)pyrene as substrate as an index of mutagenicity through the formation of reactive intermediates, the following results were obtained. With solubilized partially purified cytochromes P-450 in place of the usual liver S9 mix, there is a dose–response increase in the formation of bacterial revertants (i.e. positives for mutagenicity) with the addition of cytochrome P-448 preparations, but there is no significant dose response increase with increasing amounts of other cytochrome P-450 enzyme forms (Conney *et al.* 1977). Also, for example, the mutagenicity of protein pyrolysis products is greatly increased when activated by mouse liver microsomal fractions containing high levels of cytochrome P_1-450 (Nebert *et al.* 1979).

Cytochrome P-448 enzyme forms hydroxylate substrates at conformationally hindered positions, for example the 'bay-regions' of polycyclic aromatic hydrocarbons, and these metabolites are more likely to lead to mutagenesis. Studies on the mechanism of this effect have shown that oxygenation of carcinogenic polycyclic aromatic hydrocarbons at the bay-regions may form epoxides which are not acceptable substrates for subsequent detoxication by epoxide hydrase and other enzymes (Levin *et al.* 1977). As a result, these bay-region epoxide compounds react readily with DNA and other cellular molecules and are recognized to be the ultimate carcinogenic molecular species. It has been suggested that this may be a general phenomenon and that the carcinogenicity of environmental chemicals (xenobiotics) is often due to their activation by oxygenation in hindered positions (Parke & Ioannides 1982).

The best characterized example of metabolic activation by cytochrome P-448 is for the polycyclic aromatic hydrocarbon, benzo(a)pyrene. Sims *et al.* (1974) elucidated the pathway for the metabolic activation of benzo(a)-pyrene to the ultimate carcinogenic species benzo(a)pyrene-7,8-diol-9,10-epoxide. This compound is formed firstly through the cytochrome P-448 catalysed epoxidation of benzo(a)pyrene to benzo(a)pyrene-7,8-oxide followed by conversion by epoxide hydrase to *trans*-benzo(a)pyrene-7,8-dihydrodiol. This then undergoes further oxygenation by the cytochrome P-448 enzyme to the ultimate carcinogen. This can then react with DNA and trigger off the process of carcinogenicity by molecular events which are now beginning to be understood in terms of genomic rearrangement and promoter activation of oncogenes.

Cytochrome P-488 enzymes are not the only cytochromes P-450 responsible for carcinogen activation, and the situation may vary for different carcinogens. For example, in a study using specific antibodies to phenobarbital-induced cytochrome P-450 and 3-methylcholanthrene-induced cytochrome P-448, Kawajiri *et al.* (1983) found that although benzo(a)pyrene, 3-methylcholanthrene, 2-acetylaminofluorene, and *o*-aminoazotulene were activated selectively by 3-methylcholanthrene induced cytochrome P-448, aflatoxin B_1 was activated selectively by phenobarbital-induced cytochrome P-450.

Cytochrome P-448 activity can be measured specifically by using the substrate ethoxyresorufin which is dealkylated to resorufin selectively by cytochrome P-448 enzyme forms (Phillipson *et al.* 1984). Studies using this substrate and others, and inhibitors of cytochrome P-448 such as 9-hydroxyellipticine, have enabled some computergraphic modelling studies of the cytochrome P-448 binding site. These reveal that it is substantially different from that of less specific cytochrome P-450 enzymes such as the major isoenzyme induced by phenobarbital (Parke *et al.* 1985). Parke *et al.* (1985) have also suggested that the overall conformation of the cytochrome P-448 binding site approximates to the shape of cyclopentanophenanthrene and is similar to cholesterol and some steroid hormones, and thus this enzyme may be very closely related to other cytochrome P-450 enzymes involved in cholesterol and steroid metabolism. Several hydroxylations involved in steroid hormone biosynthesis have been shown to be mediated by cytochrome P-450 enzymes with similar properties to cytochrome P-448 enzymes (Bostrom 1983).

5.1.5 Cytochrome P-450 in prokaryotic microorganisms

Cytochrome P-450 has been detected in several species of bacteria with varying roles for the enzymes reported in each case. These are all soluble enzymes which appear to have a narrow specificity. By far the best characterized of these enzymes is that from *Pseudomonas putida*. This organism produces a cytochrome P-450 enzyme when grown on D(+)-camphor, which metabolizes camphor, enabling growth of the organism on this compound as sole carbon and energy source (Yu *et al.* 1974). This enzyme (termed cytochrome P-450_{cam}) hydroxylates the 5-methylene carbon of camphor to form the exo-5-alcohol (Gunsalus *et al.* 1975). Studies on cytochrome P-450_{cam} have been greatly facilitated because unlike mammalian cytochrome P-450 enzymes it is a soluble enzyme, i.e. it is not membrane bound. Cytochrome P-450_{cam} has been purified to homogeneity and has been crystallized. It contains 414 amino acid residues, has a molecular weight of 45 000, and a Soret peak in the reduced carbon monoxide difference spectrum at 446 nm (O'Keefe *et al.* 1978). The crystallized enzyme has now been subject to X-ray analysis and the crystal structure determined at 2.6 Å resolution (Poulos *et al.* 1985). This structure determination has revealed that the substrate molecule is buried in an internal pocket just above the haem distal surface adjacent to the oxygen binding site as might be expected. The substrate molecule is held in place by a hydrogen bond between the side-chain hydroxyl group of Tyr 96 and the camphor carbonyl oxygen atom. In addition there are some hydrophobic interactions between the substrate and neighbouring residues.

Cytochrome P-450_{cam} operates *via* an electron transport chain which contains an extra protein component known as putidaredoxin, which is a small iron-sulphur protein. Also the flavoprotein reductase contains only one flavin group, an FAD (Dus 1975). In these respects this system differs from the microsomal cytochrome P-450 systems already described, but is similar to mammalian mitochondrial cytochrome P-450 systems which also contain an extra iron-sulphur protein and use a reductase enzyme with one flavin group (Suhara *et al.* 1978). Dus *et al.* (1980) have compared the immunochemical properties of cytochrome P-450_{cam} with cytochrome P-450_{scc}, the mitochondrial cytochrome P-450 found in large amounts in the adrenal cortex of mammals that is responsible for the side-chain cleavage of cholesterol. Antibodies to one of these proteins could inhibit the activity of the other protein (enzyme), and the two antibodies showed a high degree (approximately 75%) of cross-reactivity, indicating one or more antigenic determinants in common. In these experiments, cytochrome P-450 LM_2 from phenobarbital induced rabbit liver microsomes showed 60% cross-reactivity to cytochrome P-450_{cam}.

Two cytochrome P-450 species with different specificities have been reported in different strains of *Bacillus megaterium*. The strain ATCC 14581 produces a cytochrome P-450 with *w*-2 hydroxylase activity towards fatty acids (with some *w*-1 and *w*-3 activity) which can also hydroxylate corresponding amides and alcohols (Matson *et al.* 1977). This enzyme can also catalyze the epoxidation of unsaturated fatty acids (Reuttinger & Fulco 1981). Unlike cytochrome P-450_{cam} from *P. putida*, this cytochrome P-450 enzyme from *B. megaterium* ATCC 14581 was not inducible by substrates of the enzyme, but was inducible 28-fold by phenobarbital for both cytochrome P-450 level and monooxygenase activity (Narhi & Fulco 1982). This induction was more marked when the phenobarbital was autoclaved in the culture medium, and this effect has been ascribed to the inducing effect of the hydrolytic product of phenobarbital, 2-phenylbutyrylurea (Reuttinger *et al.* 1984).

Bacillus megaterium ATCC 13368 also contains a cytochrome P-450 monooxygenase system. This system has very different properties from those of ATCC 14581 described above. The enzyme from the ATCC 13368 strain is mainly a 15β-hydroxylase (with some 6β-hydroxylase activity) for 3-oxo-Δ^4-steroids such as progesterone (Berg *et al.* 1977). This cytochrome P-450 (termed cytochrome P-450_{meg}) is a soluble enzyme and has been purified to homogeneity and found to have a molecular weight of 52 000 and an amino acid composition similar to that of cytochrome P-450_{cam} (Berg *et al.* 1979). The reduced carbon monoxide difference spectrum reveals a peak at 450 nm. This system also contains a three-protein component electron transport chain similar to that in *P. putida*, as do all the bacterial cytochrome P-450 monooxygenase systems investigated so far. One difference in the cytochrome P-450_{meg} system is that the reductase contains FMN and not FAD as the prosthetic group (Gustafsson *et al.* 1980). Studies on the small iron-sulphur protein (ferredoxin) component have revealed that its molecular weight and amino acid composition closely resemble those of putidaredoxin and adrenodoxin, the corresponding proteins in *P. putida* and mammalian adrenal cortex mitochondria (Berg 1982). In common with other bacterial systems the substrate specificity of cytochrome P-450_{meg} is very narrow, with only 3-oxo Δ^4-steroids acting as substrates (Berg & Rafter 1981). In addition, these workers found that no induction of cytochrome P-450_{meg} could be achieved with either substrates, or classical inducers, of mammalian cytochromes P-450 such as phenobarbital.

A cytochrome P-450 system has been detected in *Corynebacterium* grown on *n*-octane, which hydroxylates *n*-octane, and allows growth of the organism on *n*-octane as sole carbon and energy source (Cardini & Jurtshuk 1970). This enzyme can also oxygenate several other hydrocarbon substrates, such as benzene, toluene, and various aliphatic hydrocarbons. The enzyme is induced by its substrate (as is cytochrome P-450_{cam}), and is not found when the organism is grown on non-hydrocarbon substrates. Asperger *et al.* (1981) have shown that several strains of *Acinetobacter* contain cytochrome P-450 when grown on *n*-alkanes. These cytochrome P-450 enzymes are thought to be involved in alkane hydroxylation as the first step in their complete oxidation (Asperger *et al.* 1981). A similar hydrocarbon-inducible cytochrome P-450-dependent alkane hydroxylase is also observed in several species of yeast (see section 5.1.6).

Broadbent & Cartwright (1974) have purified and characterized a soluble cytochrome P-450 from a *Nocardia* species grown on iso-vanillate, which can *O*-dealkylate *p*-alkylphenyl ethers. This enzyme (termed cytochrome P-450_{npd}) has a molecular weight of 42–45 000 and has many properties in common with the cytochrome P-450_{cam} enzyme, although a major difference is that cytochrome P-450_{npd} does not appear to undergo a low to high-spin transition upon binding of substrate (Broadbent & Cartwright 1974).

Appleby (1978) purified three cytochrome P-450 enzymes from *Rhizobium japonicum* grown symbiotically on soybean root nodules. The three enzymes designated cytochrome P-450_{a}, $_{b}$, and $_{c}$ had Soret peaks in the reduced carbon monoxide difference spectrum at 449 nm, 448 nm, and 447 nm for cytochromes P-450_{a}, $_{b}$, and $_{c}$ respectively. The role of these *R. japonicum* enzymes is unknown, but they may have a role in the removal of oxygen in this strictly anaerobic system, or in electron transport at very low oxygen tension to generate ATP required for nitrogenase activity (Bergersen & Turner 1975).

Edelson & McMullen (1977) examined *Escherichia coli* for cytochrome P-450 activity. This organism contained an enzyme, inhibited by carbon monoxide which could dealkylate *p*-nitroanisole. This activity could be induced by phenobarbital.

Cytochrome P-450 has been reported to be involved in the bacterial luminescence of *Photobacterium fischeri* (Baranova *et al.* 1979). The luciferase complex of this organism is thought to consist of four protein components in a multienzyme complex, one of which is a cytochrome P-450. The role of the cytochrome P-450 is thought to be in the hydroxylation of aliphatic aldehydes, substrates of the luminescence reaction. Danilov *et al.*

(1982) have shown that the cytochrome P-450 from *Photobacterium fischeri* gives a type I binding spectrum with camphor, and that this compound can inhibit the luminescence system competitively.

In all of the bacterial cytochrome P-450 systems now characterized the cytochrome P-450 is soluble and operates *via* an electron transport chain of three protein components. Thus the bacterial systems are similar to those of mammalian mitochondria, but unlike those of mammalian microsomes. In addition, bacterial cytochrome P-450 systems have very narrow substrate specificities.

5.1.6 Cytochrome P-450 enzymes in eukaryotic microorganisms

Cytochrome P-450 enzymes have been demonstrated in several eukaryotic microorganisms. In contrast to bacterial systems the electron transport chain is similar to that of mammalian microsomal systems, with only two protein components: a reductase with two flavin groups (one FAD and one FMN) and a cytochrome P-450. Eukaryotic microorganisms have been found to contain their cytochrome P-450 systems bound to endoplasmic reticulum as in mammalian tissues. In general the substrate specificity of these systems appears to be broader than for bacterial systems, yet not as broad as the wide specificity cytochrome P-450 type of enzyme found in mammalian liver.

5.1.6.1 Fungi

Claviceps purpurea contains a microsomal cytochrome P-450 which is induced two-fold in both cytochrome P-450 level and activity by phenobarbital (Ambike *et al.* 1970). When treated with 3-methylcholanthrene only a small increase in the cytochrome P-450 level is observed, yet a shift in the Soret peak of the reduced carbon monoxide difference spectrum takes place form 450 nm to 448 nm. Evidence from the use of protein synthesis inhibitors suggests that this shift in wavelength is a result of *de novo* protein synthesis (Ambike *et al.* 1970), suggesting that two forms of cytochrome P-450 may occur in this organism. These workers showed a direct correlation between alkaloid production and cytochrome P-450 levels and suggested that this cytochrome P-450 enzyme is involved in alkaloid biosynthesis, although the exact reaction(s) catalyzed is unknown.

Rhizopus nigricans has been shown to contain a cytochrome P-450 enzyme which is capable of catalysing the 11α-hydroxylation of progesterone. This activity is inhibited by carbon monoxide, and this inhibition could be reversed by light at 450 nm, confirming the cytochrome P-450 nature of this activity (Breskvar & Hudnik-Plevnik 1977). Later studies showed that this enzyme system contains two protein components, the cytochrome P-450 and an NADPH : cytochrome P-450 reductase, similar to other microsomal systems. These two components have now been solubilized and partially purified (Breskvar 1983). In both *Rhizopus nigricans* and *Rhizopus arrhizus* it has been shown that the cytochrome P-450-dependent progesterone hydroxylase is induced by the substrate progesterone. There is little activity in uninduced cells (Breskvar & Hudnik-Plevnik 1981).

A cytochrome P-450 monooxygenase is present in *Cunninghamella bainieri* that is capable of *N*-demethylation of aminophenazone, *O*-demethylation of *p*-nitroanisole, and the hydroxylation of anisole, aniline, and naphthalene (Ferris *et al.* 1976). Ferris *et al.* (1976) also described an NIH shift during anisole hydroxylation, an observation supported by the previous work of Auret *et al.* (1971) which showed an 'NIH shift' of ring substituent position, during aromatic hydroxylation by nine fungal species. The 'NIH shift' is evidence for an epoxide intermediate and is well established for mammalian cytochrome P-450-catalysed aromatic hydroxylation (Jerina & Daly 1974). Neither 3-methylcholanthrene nor phenobarbital were able to induce cytochrome P-450 in *Cunninghamella bainieri*, yet phenanthrene elicited a 10-fold increase in hydroxylase activity (Ferris *et al.* 1976). Cytochrome P-450 from *C. bainieri* has been reported to be able to *N*-demethylate codeine (Gibson *et al.* 1984).

Cunninghamella elegans possesses a cytochrome P-450 capable of the hydroxylation of naphthalene which can be induced to five times its original level by naphthalene and two to three times the original level by phenobarbital or 3-methylcholanthrene (Cerniglia & Gibson 1978). This enzyme can also hydroxylate benzo(a)pyrene in a pattern very similar to that observed with some mammalian monoxygenase systems (Cerniglia & Gibson 1979). Hydroxylation of benzo(a)pyrene is inducible also in *Neurospora crassa* (Lin & Kapoor 1979) and is well established for the cytochrome P-450 system of *Saccharomyces cerevisiae* (see on). *Cunninghamella elegans* has been shown also to hydroxylate biphenyl, a well known substrate of mammalian cytochrome P-450 enzymes, although any involvement of cytochrome P-450 in this reaction is unknown (Dodge *et al.* 1979). 3-Methylcholanthrene is also metabolized by this organism (Cerniglia *et al.* 1982), which is known to also have epoxide hydrase, glutathione-S-transferase, and UDP-glucuronyltransferase activities (Wackett & Gibson 1982).

The fungus *Nectria haematococca*, a pathogen of garden pea, contains a cytochrome P-450 enzyme which can *O*-demethylate pisatin, an antimicrobial compound synthesized by infected pea tissue (Matthews & Van Etten 1983). This reaction renders the pisatin less toxic to the fungus and is inhibited by a number of cytochrome P-450 inhibitors including carbon monoxide. Carbon monoxide inhibition was reversible by light at wavelengths close to 450 nm (Matthews & Van Etten 1983). This enzyme is microsomal and requires NADPH and NADPH: cytochrome P-450 reductase for its activity (Desjardins *et al.* 1984). Multiple forms of this cytochrome P-450 may exist, because three different genes coding for the expression of pisatin demethylase have been isolated. Desjardins *et al.* (1984) have solubilized and partially purified the enzyme from pisatin-induced cultures of *Nectria haematococca* and demonstrated enzyme activity when the cytochrome P-450 was reconstituted with NADPH and NADPH-cytochrome P-450 reductase.

Cytochrome P-450 enzymes have been suggested to be involved in toxin biosynthesis by *Penicillium* (Murphy *et al.* 1974) and in fungicide degradation by *Pyricularia* (Kodama *et al.* 1982). Also several species of *Aspergillus* are capable of aromatic hydrocarbon hydroxylase activity towards benzo(a)-pyrene (Gosh *et al.* 1983) and biphenyl (Golbeck & Cox 1984).

One report has been made of a soluble fungal cytochrome P-450 from *Fusarium oxysporum* (Shoun *et al.* 1983). These workers report that several forms of cytochrome P-450 may exist in *Fusarium oxysporum* and that these were isolated as soluble enzymes, although this could be an artefact of the isolation process.

5.1.6.2 Yeasts

The occurrence of cytochrome P-450 in *Saccharomyces* yeasts is now well established. Lindenmayer & Smith (1964) were the first to demonstrate the presence of this haemoprotein in *Saccharomyces cerevisiae*, and this finding has now been extended to several species of *Saccharomyces* and related yeasts. This cytochrome P-450 system will be discussed in detail in later sections of this chapter.

Several other yeast species have been shown to contain a cytochrome P-450 enzyme when grown on *n*-alkanes as sole carbon source. These organisms use cytochrome P-450 as an alkane hydroxylase that catalyses the first step in the degradation of alkanes for growth, in a manner similar to the bacterial systems in *Corynebacterium* and *Acinetobacter* described in section 5.1.5.

The first report of an alkane grown yeast containing cytochrome P-450 was for *Candida tropicalis* grown on *n*-tetradecane as sole carbon source (Gallo *et al.* 1971). This cytochrome P-450 catalyses the *w*-hydroxylation of alkanes and of lauric acid, and is also capable of *N*-demethylation of aminopyrine, hexobarbital, benzphetamine, and ethylmorphine. This enzyme is dependent on oxygen and NADPH, is inhibited by carbon monoxide, is bound to the microsomal membrane, and requires NADPH:-cytochrome P-450 reductase and phospholipid for full activity (Duppel *et al.* 1973). Both the reductase and lipid components could be replaced by the corresponding fractions from rat liver microsomes (Duppel *et al.* 1973). Hexadecane has been shown to be a potent inducer of this system, as have all *n*-alkanes with more than 10 carbon atoms, and long chain alkenes and alcohols (Gilewicz *et al.* 1979). The components of the electron transport chain from *Candida tropicalis* have been purified and reconstituted into an active hydroxylase system (Bertrand *et al.* 1979). Mansuy *et al.* (1980) have shown that cytochrome P-450 from *Candida tropicalis* is in a spin state equilibrium similar to that of mammalian microsomal cytochrome P-450, and that this enzyme can also undergo spectral changes on binding of ligands.

A similar enzyme has been described in *Candida guilliermondii* grown on *n*-alkanes, which hydroxylates long chain alkanes (preferably hexadecane to octadecane) to their primary alcohols (Muller *et al.* 1979). This microsomal enzyme also produces binding spectra with several compounds, including a type I spectrum with the substrate hexadecane. It was shown that *C. guilliermondii* could not produce cytochrome P-450 during growth on glucose, nor when glucose was added together with the hydrocarbon inducer

(Muller *et al.* 1979). Mauersberger *et al.* (1980) have shown that the level of alkane-induced cytochrome P-450 in *C. guilliermondii* is higher when the oxygen level is low. These workers suggested that this effect might be due to oxygen limitation causing a decreased alkane hydroxylation rate, to which the yeast responds by increasing biosynthesis of the monooxygenase.

Mauersberger & Matyashova (1980) reported the occurrence of cytochrome P-450 also in *Candida lipolytica* grown on *n*-alkanes, and Il'chenko *et al.* (1980) demonstrated a similar system in *Torulopsis candida*. *Lodderomyces elongisporus* (*Candida maltosa*) is also capable of producing a cytochrome P-450 when grown on *n*-alkanes, particularly tetradecane (Mauersberger *et al.* 1981). This enzyme and its reductase have been purified and reconstituted into an active hydroxylase system, which requires phospholipid (Muller *et al.* 1982). The level of this cytochrome P-450 is greatly increased by growing at low oxygen concentration (Mauersberger *et al.* 1984), and changes in enzyme levels during the switch from glucose to *n*-alkane growth are controlled at the transcriptional level (Sunairi *et al.* 1984). *Saccharomycopsis lipolytica* also produces a cytochrome P-450 alkane hydroxylase when grown on *n*-alkanes which is very active towards lauric acid hydroxylation (Marchal *et al.* 1982).

Sanglard *et al.* (1984) have suggested that two distinct types of cytochrome P-450 enzymes can be detected in *Candida tropicalis* under different growth conditions. When grown on *n*-alkanes the alkane hydroxylase system described above is produced, with an enzyme with a peak in the reduced carbon monoxide difference spectrum at 450 nm. However, when grown on glucose, a different cytochrome P-450 is produced with a peak at 448 nm in the reduced carbon monoxide difference spectrum (similar to the enzyme found in *Saccharomyces* yeasts grown on glucose described in later sections of this chapter). This cytochrome P-448 enzyme had *N*-demethylase activity towards aminopyrine. Levels of both of these cytochrome P-450 enzymes were increased at low oxygen concentrations under their respective conditions. No alkane hydroxylase cytochrome P-450 was found in the absence of alkane induction, but it is possible that alkane grown cells contain both *C. tropicalis* cytochrome P-450 systems and this might account for the *N*-demethylase activity found in these cells.

5.2 CYTOCHROMES P-450 IN *SACCHAROMYCES CEREVISIAE*

5.2.1 The biosynthesis of cytochromes P-450 in *S. cerevisiae*

After the presence of a cytochrome P-450 enzyme in *Saccharomyces cerevisiae* was first demonstrated by Lindenmayer & Smith (1964), several studies have been carried out in order to investigate the conditions required for its biosynthesis. In the first study by Lindenmayer & Smith (1964) it was found that high concentrations of a cytochrome P-450 accumulated in *S. cerevisiae* grown anaerobically in a medium containing 4% (w/v) glucose, and much lower concentrations were present when the yeast was grown aerobically in the same medium.

The production of cytochrome P-450 in *S. cerevisiae* under semi-anaerobic growth conditions was confirmed by Ishidate *et al.* (1969a), with only very little cytochrome P-450 (approx. 10% of the semi-anaerobic value) produced under fully aerobic conditions. During growth at low glucose concentration (1% w/v), no cytochrome P-450 was observed. Ishidate *et al.* (1969a) demonstrated also that this enzyme was located in the microsomal fraction of the cell, and that its spectral properties were similar to those of mammalian microsomal cytochrome P-450 enzymes. These finding were later confirmed by Yoshida *et al.* (1974a). Ishidate *et al.* (1969a) found also that cytochrome P-450 was present in yeast grown semi-anaerobically but was lost on exposure to aerobic conditions. This loss paralleled the formation of active mitochondria within the yeast cells and could be prevented by high concentrations of glucose or chloramphenicol, both of which also prevented the development of mitochondrial respiration. Essentially similar findings were made by Rogers & Stewart (1973), who found a maximum cytochrome P-450 level in yeast grown in a medium containing 4% (w/v) glucose at the low oxygen concentration of 0.25 μM. The cytochrome P-450 level declined at both lower and higher oxygen concentrations. No cytochrome P-450 could be detected in yeast grown under strictly anaerobic conditions.

Subsequent work has established that cytochrome P-450 production in *S. cerevisiae* is associated with the exponential growth phase under conditions of fermentative growth such that mitochondrial cytochromes are repressed.

Following these early studies, work on cytochrome P-450 from *S. cerevisiae* has been done largely on the enzyme from yeast grown under two sets of conditions. Yoshida *et al.* (1972, 1975a, b, 1977) have worked on the enzyme produced form a wild-type yeast grown semi-anaerobically in a medium containing 6% (w/v) glucose, whereas Wiseman *et al.* (1975, 1976, 1978) have worked on the enzyme from a brewers yeast strain grown aerobically in 20% (w/v) glucose medium. Both of these sets of conditions achieve repression of mitochondrial cytochromes and the accumulation of a high level of cytochrome P-450, but these are probably not the same isoenzyme form.

Cytochrome P-450 has been shown to be produced in several other species of *Saccharomyces* and related yeasts. Cartledge *et al.* (1972) demonstrated a cytochrome P-450 in *Saccharomyces carlsbergensis* (now renamed *S. uvarum*) which was rapidly lost on adaptation from fermentative to respiratory growth conditions. Sauer *et al.* (1982) have worked on a cytochrome P-450 produced by a strain of *S. uvarum*. Poole *et al.* (1974) demonstrated the presence of a cytochrome P-450 in the fission yeast *Schizosaccharomyces pombe*. Karenlampi *et al.* (1980) scanned a number of yeast species for their ability to produce cytochrome P-450 after growth in 5% (w/v) glucose mediun. Of the 16 species tested, cytochrome P-450 was detected in 13 after growth on glucose, including: *Saccharomyces bayanus*, *Saccharomyces chevalieri*, *Schizosaccharomyces japonicum*, *Pichia fermen-*

tans, *Debaromyces hansenii*, *Hansenula anomala*, *Kluyveromyces fragilis*, *Torulopsis dattila*, *Torulopsis glabrata*, and particularly high levels in *Brettanomyces anomalus*. The high level of cytochrome P-450 found in *Brettanomyces anomalus* (approx. 0.5% of total cellular protein) when grown in 5% glucose medium, facilitated further studies on the purification of this enzyme by polymer phase partition (Karenlampi *et al*. 1986).

Karenlampi *et al*. (1980) also demonstrated that a cytochrome P-450 was present in strains of *Saccharomyces italicus* at higher levels than in *S. cerevisiae*. Three *S. italicus* strains (now renamed as strains of *S. cerevisiae*) were examined by King (1982) and compared to other strains of *S. cerevisiae*. Higher levels of cytochrome P-450 in the *S. italicus* strains were not observed.

Karenlampi *et al*. (1980) could detect cytochrome P-450 in *Candida tropicalis* only after growth on *n*-alkanes, and not after growth on glucose. These results are at variance with those of Sanglard *et al*. (1984) who also looked at cytochrome P-450 from *C. tropicalis* grown on both alkanes and glucose. These workers suggested that cytochrome P-450 was present in *C. tropicalis* under both sets of conditions, but during growth on glucose the interfering absorbance of cytochrome oxidase makes it difficult to detect cytochrome P-450 in whole cells.

The amount of cytochrome P-450 produced during growth on high glucose containing medium by strains of *S. cerevisiae* is highly strain-dependent. King *et al*. (1983a) surveyed 18 haploid strains of *Saccharomyces cerevisiae*, finding large differences in the levels of cytochrome P-450 produced (assayed in whole cells). Qureshi *et al*. (1980) reported that the brewing strain of *S. cerevisiae* N.C.Y.C. No. 239 gave higher levels of cytochrome P-450 than the strain N.C.Y.C. No. 240 used for much previous work although cytochrome P-450 could not be detected in cells of several other yeast strains used.

The *S. cerevisiae* brewing strain N.C.Y.C. No. 240 was identified as consisting of two morphological variants, N.C.Y.C. No. 753 and N.C.Y.C. No. 754 (Dr Barbara Kirsop, personal communication). The variant N.C.Y.C. No. 754 is the major (80%) component, and N.C.Y.C. No. 753 is the minor (20%) variant. These strains were examined for their ability to produce cytochrome P-450 detected spectrally in whole cells and microsomal fraction and for their ability to catalyse the cytochrome P-450-dependent reaction, benzo(a)pyrene hydroxylation (King *et al*. 1983b). The major morphological variant, N.C.Y.C. No. 754, produced a far higher level of cytochrome P-450 (8 nmol/g wet weight of yeast) than N.C.Y.C. No. 240 (3 nmol/g wet weight of yeast). Little difference was seen in the time taken to reach the peak value which was at the end of the exponential growth phase in each case. The minor variant, N.C.Y.C. No. 753, produced only a small amount of cytochrome P-450 (less than 1 nmol/g wet weight of yeast). Therefore, as expected, the N.C.Y.C. No. 240 produced an intermediate level of the enzyme, but as N.C.Y.C. No. 240 consists of 80% N.C.Y.C. No. 754, the marked differences between these two yeast strains is surprising.

The specific activity of the cytochrome P-450 from all of these strains towards benzo(a)pyrene hydroxylation was identical (King *et al.* 1983b).

Cytochrome P-450 production in *S. cerevisiae* is rapid during the exponential phase of growth but ceases at the beginning of stationary phase (Woods 1979, Karenlampi *et al.* 1981). During stationary phase the enzyme levels begin to decline, as cytochrome P-450 appears to be degraded in an energy and oxygen-requiring process that also needs mitochondrial protein biosynthesis (Blatiak *et al.* 1980). It has been shown that tween 80 is capable of increasing the level of cytochrome P-450 in yeast grown into stationary phase, presumably owing to prevention of degradation rather than to increased biosynthesis after this time (Wiseman *et al.* 1976).

The yield of cytochrome P-450 from *S. cerevisiae* grown in complex medium in batch culture has been optimized in shake flask (Salihon *et al.* 1983) and in batch fermentations (Salihon *et al.* 1985). These experiments resulted in an increased yield of cytochrome P-450 by 255% (nmol/g dry weight of yeast) and 173% in enzyme productivity (nmol/litre of culture medium). The most significant factors in enzyme yield and productivity were found to be impeller speed and air flow-rate (in 14% glucose-containing medium). The effect of agitation was principally in maintaining optimal dissolved oxygen availability (Salihon *et al.* 1985).

Cyclic AMP may be involved in the regulation of cytochrome P-450 in *S. cerevisiae* (Wiseman *et al.* 1978). The addition of cyclic AMP to yeast spheroplasts greatly reduces the production of cytochrome P-450 during subsequent growth in 5% glucose medium, whereas 5′-AMP and 2′-(3′)-AMP have no effect (Wiseman *et al.* 1978). Wiseman *et al.* (1978) thus suggested that cyclic AMP may be involved in the control of cytochrome P-450 accumulation by a form of repression. As might be expected, cytochrome $a+a_3$ and cyclic AMP show a direct relationship to each other and an inverse relationship to the cytochrome P-450 level at various glucose concentrations (Qureshi *et al.* 1980). An inverse relationship between cytochrome P-450 and cytochrome $a+a_3$ has also been shown in yeast grown semi-anaerobically (Schunck *et al.* 1978). It has more recently been shown that cytochrome P-450 from rabbit liver microsomes is phosphorylated by a cyclic AMP-dependent protein kinase at a site conserved through evolution (Müller *et al.* 1985). This phosphorylation resulted in a decrease in cytochrome P-450 level and conversion into cytochrome P-420 followed by haem loss. If a similar mechanism exists in *S. cerevisiae* it could explain the rapid loss of cytochrome P-450 seen in stationary phase cells.

The production of cytochrome P-450 during growth of *S. cerevisiae* on a range of sugars has been studied by Karenlampi *et al.* (1981a). These workers found that high levels of cytochrome P-450 were accumulated during growth of yeast at high concentrations of a strongly fermentable sugar such as glucose, fructose or sucrose. Cytochrome P-450 was also produced, although to a lesser extent, during growth on galactose or maltose, where fermentation and respiration occurred concomitantly. However, when a non-fermentable carbon source was used for growth (such as glycerol, lactate, or ethanol) no cytochrome P-450 was produced. These workers thus associated cytochrome P-450 synthesis with conditions of rapid growth and fermentation, though not necessarily to repression of mitochondria. Because the required haem is synthesized during both fermentative and respirative conditions (Labbe-Bois & Volland 1977), it is probable that apoprotein synthesis is regulated in these different conditions.

One study of the genetic regulation of cytochrome P-450 production in *S. cerevisiae* has been carried out (King *et al.* 1983a). A cross was performed between two haploid yeast strains, one of which produced a high level of cytochrome P-450 and one of which produced no spectrally detectable cytochrome P-450. The diploid and several meoitic tetrads were then examined for their cytochrome P-450 levels. No spectrally detectable cytochrome P-450 was found in the resulting diploid, and in the meiotic tetrads a 2:2 segregation of cytochrome P-450 producer and non-producer strains was observed. Therefore a single nuclear gene appears to control the production of cytochrome P-450. However, because the diploid produced no cytochrome P-450 this single nuclear gene is probably not the structural gene, but a regulatory gene, for if all that was necessary for cytochrome P-450 production was the possession of a functional allele of the structural gene then the heterozygous diploid would have produced cytochrome P-450. Of the two segregants which produced cytochrome P-450 in each tetrad, one produced a level similar to the parent producer strain, and the other produced approximately twice as much. Thus the regulatory gene controlling cytochrome P-450 production appears to be modified by a second nuclear gene which increases the level of cytochrome P-450. These results led to the proposal of a simple model for the genetic control of cytochrome P-450 production in *S. cerevisiae* as shown in Fig. 5.3.

In this model (Fig. 5.3) both parent strains (and thus all of the tetrad segregants) possess a structural gene for cytochrome P-450 (cyt), although this is not expressed in one parent which contains the dominant rep^+ allele of a regulatory gene (rep) and is thus a non-producer. In the diploid, the dominance of rep^+ must prevent cytochrome P-450 production, and in the tetrads the 2:2 segregation of producers to non-producers reflects the segregation of the rep^+/rep^- alleles. The differences in cytochrome P-450 levels in the producer strains are regulated by the modifier gene (mod) inherited from the non-producing parent to increase the level of cytochrome P-450 in producer strains. The identities of such postulated regulatory and modifier genes are unknown at present.

In our laboratory, T.-K. Lim (1976) reported on the production of cytochrome P-450 in yeast grown in continuous culture. More recently, Trinn *et al.* (1982) studied the occurrence of cytochrome P-450 in *S. cerevisiae* during growth in continuous culture. All of these workers showed that glucose and oxygen are major regulatory effectors for cytochrome P-450 production, similarly to when grown in batch culture. For cytochrome P-450 production both glucose repression and oxygen limitation were required. Trinn *et al.* (1982) reported that under glucose-derepressed

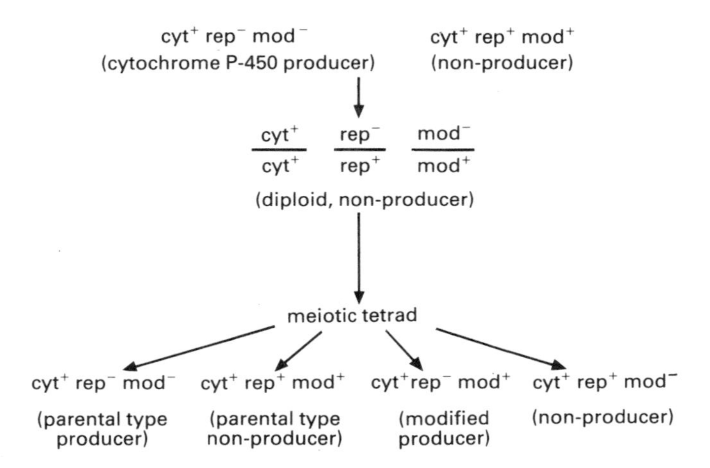

Fig. 5.3 — Model of the genetic control of cytochrome P-450 production in *Saccharomyces cerevisiae*. cyt^+: structural gene for cytochrome P-450. rep^-/rep^+: non-repressing and permanently repressing alleles respectively of a regulatory gene. rep^+ is dominant in the heterozygote. mod^-/mod^+: alleles of a modifying gene active in rep^- cells. The presence of mod^+ enhances the amount of cytochrome P-450 produced and is an important part of the proposed model.

conditions it was not possible to induce the formation of cytochrome P-450 by oxygen limitation alone. Under glucose-repressed conditions high levels of cytochrome P-450 were produced below a dissolved oxygen tension of approximately 15% (Trinn *et al.* 1982).

The influence of oxygen on cytochrome P-450 accumulation was investigated also by Bertrand *et al.* (1984). These workers demonstrated that cytochrome P-450 levels were high when *S. cerevisiae* was grown in 3% glucose medium at low oxygen tensions. This effect was seen also for NADPH: cytochrome P-450 reductase, cytochrome b_5 and NADH: cytochrome b_5 reductase, which all increased at low oxygen levels. Also the membrane surface of the endoplasmic reticulum appeared to increase when the yeast was grown at low oxygen tension.

The importance of oxygen in the production of cytochrome P-450 in *S. cerevisiae* was again emphasized by Blatiak *et al.* (1985a). *S. cerevisiae* grown in 20% glucose medium accumulated cytochrome P-450 up to a maximum level at the end of the exponential growth phase. The accumulation of cytochrome P-450 was associated with rapid growth, with the enzyme level declining in stationary phase. The effect of growing cultures at different shake speeds was examined in this study. In all cases the highest level of cytochrome P-450 was seen near the end of exponential growth, although the growth rates were of course different in each case. The optimum shake speed for growth (and total biomass) was also the optimum for cytochrome P-450 accumulation. This optimum shake speed was thought to reflect the optimum degree of aeration of the culture for cytochrome P-450 production, although other effects such as homogeneity of the shake culture could not be ruled out. In parallel work using a 5-litre stirred tank fermenter, agitator speed and air-flow rate into the culture were found to affect the yield of cytochrome P-450 from *S. cerevisiae* very strongly, the results being significantly different at a 99% confidence level (Blatiak *et al.* 1985a). Culture conditions were optimized with respect to a number of operational parameters including agitator speed and air-flow rate, producing a yield of 126 nmolg^{-1} (dry weight of yeast). When optimum conditions were maintained, except that instead of using the optimum air-flow rate of 9 litres h^{-1}, a gas flow consisting of 6 litres h^{-1} air mixed with 3 litres h^{-1} oxygen was used (i.e. fermentation conditions were identical except for an increased dissolved oxygen tension), the enzyme yield was reduced to 34.5 nmolg^{-1} (dry weight of yeast), although the biomass yield was constant. In another experiment where dissolved oxygen tension was reduced (by having a gas flow of 6 litres h^{-1} air and 3 litres h^{-1} nitrogen) the enzyme yield was again reduced, this time to 30.4 nmolg^{-1} (dry weight of yeast), again with a similar biomass. Thus the effect of the optimum aeration and agitation conditions was to maintain optimum dissolved oxygen availability.

Blatiak *et al.* (1985a) claimed that at low oxygen concentrations oxygen acts as an inducer of cytochrome P-450. This was shown by making yeast cultures anaerobic in early exponential growth, which resulted in the prevention of cytochrome P-450 production without significantly altering the yeast growth rate. The role of oxygen may be as a substrate inducer of yeast cytochrome P-450, the removal of oxygen thus slowing the induction of this enzyme. When cultures of *S. cerevisiae* were made anaerobic during early stationary phase, cytochrome P-450 was not lost as quickly as in the aerobic control. At stationary phase the cytochrome P-450 level is already falling, and thus removal of oxygen at this time is unlikely to affect cytochrome P-450 biosynthesis. The slowing of the loss of cytochrome P-450 is likely to be due to protection from degradation initiated by oxygen as indicated by Blatiak *et al.* (1980).

An endogenous role of cytochrome P-450 in *S. cerevisiae* is thought to be in ergosterol biosynthesis (see on). Under strictly anaerobic conditions *S. cerevisiae* requires added ergosterol and a fatty acid source for growth, because oxygen is required for their biosynthesis. Therefore, under these anaerobic conditions, cytochrome P-450 is not needed for sterol biosynthesis and does not appear to be produced (Rogers & Stewart 1973). When oxygen is introduced in small quantities, cytochrome P-450 is induced, presumably for use in sterol biosynthesis. However, at very high oxygen levels the cytochrome P-450 level declines.

Ethanol might also be important in the regulation of cytochrome P-450 in *S. cerevisiae*. Morita & Mifuchi (1984) found that the levels of cytochrome P-450 were increased when the growth medium (containing 2% glucose) was supplemented with 1.5% or 3% ethanol. This induction was not seen when cells were grown in medium supplemented with 6% ethanol, nor when grown on ethanol as sole carbon source. As ethanol is of course produced in large quantities during the fermentation of glucose, Del Carratore *et al.* (1984) examined the hypothesis that it was ethanol that was inducing the high levels of cytochrome P-450, rather than the glucose. At glucose levels higher than 3% the supplementation of the medium with ethanol resulted in the prevention of cytochrome P-450 production. However, during growth at low glucose concentration (0.5%) where no cytochrome P-450 could normally be detected spectrally, the addition of 2% ethanol resulted in the induction of a small amount of cytochrome P-450.

The induction of cytochrome P-450 by the addition of 1% ethanol into the medium of *S. cerevisiae* growing in 0.5% glucose medium has been reported by Blatiak *et al.* (1985b, 1987). These workers also showed that added ethanol could marginally prevent the loss of cytochrome P-450 in non-growing yeast, possibly by inducing some new cytochrome P-450 production. However, at present it is not known whether the enzyme induced by *added* ethanol is the same as that induced during growth at high glucose concentration (where ethanol is formed *inside* the yeast cell), or whether a different form of the enzyme is induced. The protective effects of ethanol seem to be specific, as other alkanols cause rapid loss of cytochrome P-450 in the yeast, the rate of loss being directly related to the lipid solubility of the alkanol (Blatiak *et al.* 1985b, 1987). Ethanol is a well known inducer of a mammalian cytochrome P-450 in hepatic tissue which operates *via* a free-radical mediated mechanism (Ingelman-Sundberg & Hagbjork 1982). The ethanol-induced yeast enzyme may be similar to this mammalian ethanol-induced iosoenzyme and may also operate *via* a free-radical mediated mechanism. It is interesting to note that the mammalian ethanol-induced enzyme has a very high activity for the oxidation of ethanol to acetaldehyde (Coon *et al.* 1984), and the yeast enzyme may be involved in a similar role during fermentation at high glucose concentration when large amounts of ethanol are produced. Such an enzymic role for yeast cytochrome P-450 would be very important in explaining and predicting the ethanol-tolerance of yeasts in the presence of high concentrations of ethanol produced by the large amounts of glycolytic enzymes present in the yeast under these conditions of rapid growth.

The metabolisn of 8-methoxypsoralen by *S. cerevisiae* is also induced by ethanol (Prognon *et al.* 1984), and it is possible that this activity is cytochrome P-450-dependent.

It is known that the production of the yeast haemoproteins iso-1-cytochrome c, catalase T, and catalase A are coordinately controlled by glucose, oxygen, and haem through the control of messenger RNA levels (Hortner *et al.* 1982), although there is also regulation at the post-transcriptional level (Hortner *et al.* 1982, Laz *et al.* 1984). The levels of iso-1-cytochrome c and iso-2-cytochrome c are reduced in yeast grown anaerobically or under glucose repression, whereas the amount of apoprotein of each of cytochrome b, cytochrome c, and cytochrome c peroxidase is not reduced (Ross & Schatz 1976, Ciejan *et al.* 1980). The production and addition of haem is thus a crucial controlling step in these systems, although it should be noted that biosynthesis of the haem moiety occurs in mitochondria (Gudenus *et al.* 1984). It has been shown recently that haem is capable of being transferred between different haemoproteins including cytochrome P-450

(Wyman *et al.* 1986). Different haemoproteins respond differently to glucose, oxygen, haem, and ethanol.

It is obvious that a complex mechanism of regulation of cytochrome P-450 is present in *S. cerevisiae*. Oxygen induces cytochrome P-450 at low levels, whereas high levels of oxygen repress or destroy the enzyme (Blatiak *et al.* 1985a, 1987). Glucose and ethanol are also involved in the regulation of this enzyme, and it is likely that haem plays a role in regulating cytochrome P-450 levels as has been shown for iso-1-cytochrome c (Guarente & Mason 1983). An interesting review has been published recently on cytochrome P-450 from yeasts (Kappeli 1986). See also Chapter 3, pp. 59–60.

5.2.2 Enzymic activities of cytochrome P-450 from *S. cerevisiae*

The presence of a complete microsomal electron transport system in *S. cerevisiae*, similar to that in liver microsomes, was demonstrated initially by Yoshida *et al.* (1974). This group have isolated and purified all of the protein components involved; cytochrome P-450, NADPH: cytochrome P-450 reductase, cytochrome b_5, and NADH: cytochrome b_5 reductase, from yeast grown semi-anaerobically (Kubota *et al.* 1977, Yoshida *et al.* 1977, Yoshida & Aoyama 1984). The cytochrome P-450 component was shown to be a cytochrome P-448 enzyme with a reduced-carbon monoxide spectral peak at 447–448 nm. Initially a molecular weight of 51 000 was reported for this cytochrome P-448 (Yoshida *et al.* 1977), although more recently this has been corrected to a value of 58 000 (Yoshida & Aoyama 1984). In intact yeast the cytochrome P-450 was thought to be largely in a high spin state and thus substrate bound (Yoshida & Kumaska 1975a,b). However, when isolated and purified the enzyme is found largely in the low spin form, and detergents used in the purification procedure may be responsible for this (Yoshida & Aoyama 1984). This was also shown to be the case by King *et al.* (1984). A cytochrome P-450 enzyme was isolated from microsomes prepared from yeast grown in 20% (w/v) glucose medium, and purified to homogeneity, 88 to 97% pure on a specific content basis (Azari & Wiseman 1982a, King *et al.* 1984). The purified enzyme had a peak in the reduced-carbon monoxide difference spectrum at 448 nm and a molecular weight of about 55 000 as determined by SDS-polyacrylamide gel electrophoresis. This enzyme was shown to be approx. 94% low spin (at 22°C) when purified, and amino acid analysis revealed a high content of hydrophobic residues (43%) as would be expected for an integral membrane protein (King *et al.* 1984).

5.2.2.1 The role of yeast cytochrome P-450 in the metabolism of endogenous substrates

The major sterol in yeast membranes is ergosterol. The pathway of ergosterol biosynthesis in yeast has many similarities to the pathway of cholesterol biosynthesis in mammalian tissue, and indeed as far as lanosterol the pathway of sterol biosynthesis is identical. In both pathways lanosterol is a key intermediate and the lanosterol 14α-demethylation reaction (removal of the C32 methyl group, see Fig. 5.4) is an essential and rate-limiting step.

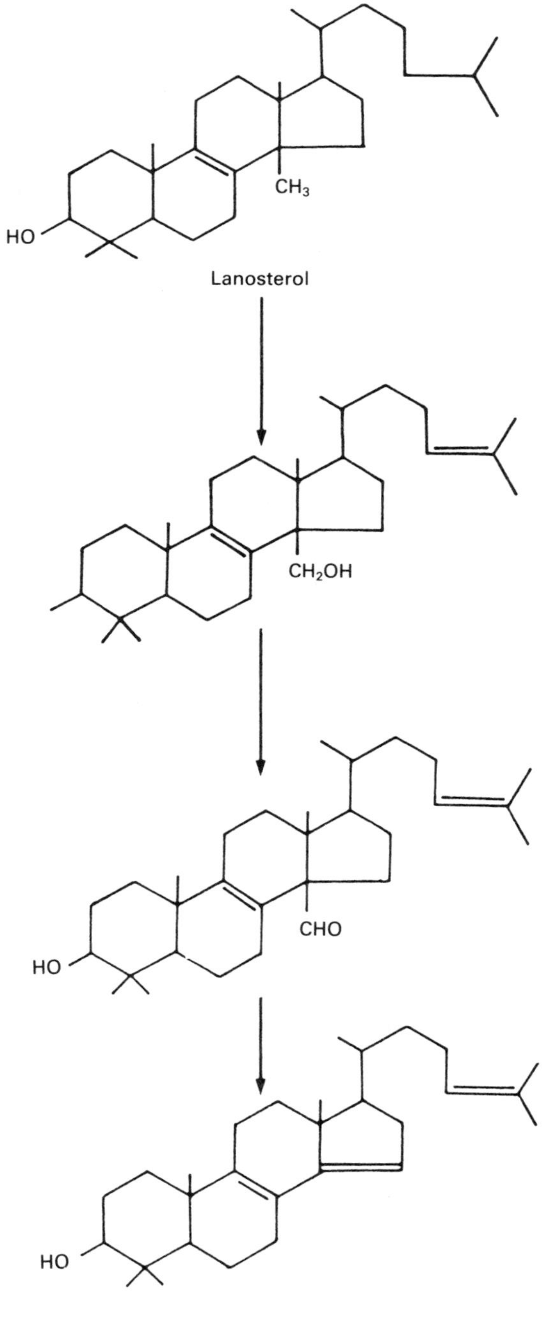

Fig. 5.4 — The 14 α-demethylation of lanosterol.

This step is thought to be catalysed by a cytochrome P-450 enzyme in both mammals (Gibbons *et al.* 1979) and yeast (Aoyama *et al.* 1984; Yoshida & Aoyama 1984). This yeast enzyme named cytochrome P-450_{4DM} does not give a binding spectrum with benzo(a)pyrene (Yoshida & Aoyama 1984) and appears to be distinct from the isoenzyme studied by King *et al.* (1984) who grew their yeast aerobically in 20% glucose, rather than semi-anaerobically at lower glucose-concentration (3%–6%).

Alexander *et al.* (1974) showed that the cytochrome P-450 inhibitor, carbon monoxide, could inhibit lanosterol 14α-demethylation by 57%. It was also shown that partially purified low spin cytochrome P-450 from *S. cerevisiae* could be partially converted to the high spin state by the addition of lanosterol (Yoshida & Kumaoka 1975a, b). Aoyama & Yoshida (1978a) demonstrated that lanosterol stimulated the oxidation of NADPH by purified yeast cytochrome P-450 and NADPH : cytochrome P-450 reductase, and found a type I binding spectrum for lanosterol with the purified yeast enzyme, indicative of a substate relationship. More direct proof of the role of yeast cytochrome P-450 in lanosterol 14α-demethylation was obtained by the reconstitution of an active lanosterol 14α-demethylase using only purified cytochrome P-450, purified NADPH : cytochrome P-450 reductase, phospholipid, and NADPH (Aoyama *et al.* 1984). This activity resulted in the formation of 4,4-dimethyl-5α-cholesta-8,14,24-trien-3β-ol (Fig. 5.4) which is then reduced by an NADPH-linked reductase to form 4,4-dimethyl-zymosterol. The lanosterol 14α-demethylase reaction was inhibited by the cytochrome P-450 inhibitors, carbon monoxide, metyrapone, and SKF 525A, and by antibodies to the purified cytochrome P-450 component (Aoyama *et al.* 1984).

The postulation of an endogenous role for yeast cytochrome P-450 posed the question of how ergosterol was synthesized under conditions of aerobic growth at low glucose concentration, when no cytochrome P-450 can be detected spectrally (see section 5.2.1). However, under these conditions some cytochrome P-450 is probably present, though at very low levels. Aoyama *et al.* (1981) have shown that yeast grown under these conditions is capable of lanosterol 14α-demethylation, and this activity is subject to inhibition by antibodies to purified yeast cytochrome P-450. The growth of yeast under fermentation conditions, which results in much higher levels of cytochrome P-450, resulted in higher levels of lanosterol 14α-demethylation (Aoyama *et al.* 1981).

The lanosterol 14α-demethylase reaction is thought to occur *via* three oxygenation steps, and cytochrome P-450 may be involved in one or all of these. They all require NADPH and molecular oxygen (Gibbons *et al.* 1979). Carbon monoxide appears to inhibit only the first of these steps (Gibbons *et al.* 1979), whilst a purified system of cytochrome P-450, NADPH : cytochrome P-450 reductase and phospholipid can convert lanosterol through all three steps of the reaction to 4,4-dimethyl-5α-cholestra-8,14,24-trien-3β-ol (Aoyama *et al.* 1984). If all three oxygenation reactions are catalysed then this reaction is formally similar to the side-chain cleavage of cholesterol to form pregnenolone by cytochrome P-450_{SCC} from adrenal cortex mitochondria (Takikawa *et al.* 1978), and the C19-demethylation of androstenedione to estrone by aromatase, a placental cytochrome P-450 (Thompson & Siteri 1974).

More recently the role of cytochrome P-450 in lanosterol 14α-demethylation has received a great deal of attention. This is because this enzyme is the target for a number of agriculturally or medically useful antifungal agents, such as ketoconazole, which is a potent orally-active antifungal agent used medically (for review see Coulson *et al.* 1984). These antifungal agents are imidazole or triazole derivatives. They cause the accumulation of C-14 methyl sterols such as lanosterol, and this disrupts membrane structure such that the membrane becomes permeable to protons (Thomas *et al.* 1983). Also, the activity of membrane-bound enzymes, such as a fatty acid desaturase, are decreased (Van den Bossche *et al.* 1983). This effect is due to the inhibition of cytochrome P-450-catalysed lanosterol 14α-demethylation, and this is probably the basis of their antifungal activity (Van den Bossche *et al.* 1980).

Ketoconazole has been shown to interact directly with yeast cytochrome P-450 *in vivo* by spectral means (Van den Bossche & Willemsens 1982), and this interaction correlates directly with the inhibition of lanosterol 14α-demethylation (Van den Bossche *et al.* 1984). Cytochrome P-450 from rat liver microsomes was much less sensitive to inhibition by ketoconazole (Van den Bossche & Willemsens 1982). The agricultural fungicide, buthiobate, has also been shown to be a potent inhibitor of yeast cytochrome P-450-catalysed lanosterol 14α-demethylation, and was shown to elicit a type II binding spectrum (Aoyama *et al.* 1983a). Wiggins & Baldwin (1984) have shown type II binding for the agricultural fungicide diclobutrazol and a number of diclobutrazol isomers. In this case the extent of binding to cytochrome P-450 was indeed shown to correlate with the antifungal activity of the compound.

Many different sterol mutants of *S. cerevisiae* have been isolated on the basis of resistance to the polyene macrolide antibiotic, nystatin. The sensitivity of yeast to nystatin is related to the level of ergosterol in the membrane (Lees *et al.* 1984), and these mutants are incapable of ergosterol biosynthesis. Several of these mutants have been shown to be deficient in lanosterol 14α-demethylation, such as Nys P-100 (Pierce *et al.* 1978) and SG1 (Trocha *et al.* 1977). SG1 has been shown to produce an inactive cytochrome P-450 enzyme (Aoyama *et al.* 1983a). This inactive cytochrome P-450 has been partially purified and shown to have spectral properties distinct from the native enzyme (Yoshida *et al.* 1985). The reduced iron, carbon monoxide difference spectra of cytochrome P-450 of a number of nystatin-resistant mutants were examined by King *et al.* (1985). In these experiments the strain Nys P-100 was shown to have a severely distorted cytochrome P-450 spectrum with a peak at 445 nm compared to 448 nm for the active enzyme. This inactive (or partially active) enzyme appeared to be overproduced. With SG1 only a small spectral difference from the active enzyme was discerned, suggesting a different distortion in the enzyme molecule. Nystatin resistance in the C-14 demethylation deficient SG1 was shown to be determined by a single gene (King *et al.* 1985).

A second role for a yeast cytochrome P-450 in ergosterol biosynthesis has been reported by Hata *et al.* (1981). These workers claimed that yeast cytochrome P-450 was also involved in the Δ^{22}-desaturation of ergosta-5,7-dien-3β-ol to form ergosterol. This role was suggested solely on the basis of studies with the cytochrome P-450 inhibitors carbon monoxide and metryapone, and thus is not definitive. This enzyme was shown to be different from the cytochrome P-450 involved in lanosterol 14α-demethylation by non cross-reaction with antibodies to purified cytochrome P-450, non-inhibition by buthiobate (a specific cytochrome P-450 inhibitor), and an increase in activity on aerobic adaptation. In addition, studies with mutants lacking a functional lanosterol demethylase contained significant levels of Δ^{22}-desaturation activity (Hata *et al.* 1983). These workers claimed therefore that the Δ^{22}-desaturase was a new form of cytochrome P-450 from yeast, although another conclusion may be that this enzyme is not cytochrome P-450-related even though it is inhibited by carbon monoxide and metryrapone.

5.2.2.2 The role of yeast cytochrome P-450 isoenzymes in the metabolism of xenobiotics (see Chapter 6.5)

The metabolism of benzo(a)pyrene by a cytochrome P-450 monooxygenase system from *S. cerevisiae* is now well established (Wiseman & Woods 1979, King *et al.* 1984), although this is at a much lower rate than in rat liver. Benzo(a)pyrene is hydroxylated by this enzyme to a range of products, the major metabolites as shown by high-performance liquid chromatography being 3-hydroxybenzo(a)pyrene, 9-hydroxybenzo(a)pyrene, and 7,8-dihydro-7,8-dihydroxybenzo(a)pyrene (Wiseman & Woods 1979). The involvement of yeast cytochrome P-450 in this activity has been demonstrated by thermal stability experiments and the inhibition of the reaction by carbon monoxide (Woods 1979). Studies using tritium NMR on substrate–haem interactions in a partially pure enzyme preparation also supported the involvement of cytochrome P-450 in benzo(a)pyrene hydroxylation (Libor *et al.* 1980). The involvement of cytochrome P-450 in benzo(a)pyrene hydroxylation was established unequivocally by the reconstitution of an active benzo(a)pyrene hydroxylase using purified yeast cytochrome P-450, purified NADPH: cytochrome P-450 reductase, and phospholipid (Azari & Wiseman 1982a, b, King *et al.* 1984). The omission of cytochrome P-450, NADPH: cytochrome P-450 reductase, or NADPH abolishes activity, and the omission of phospholipid results in approximately half of maximal activity. The phospholipid requirement can also be fulfilled by the non-ionic detergent emulgen 911, perhaps suggesting a structural role for phospholipid in associating the two proteins together (Azari & Wiseman 1982a, b). This cytochrome P-450 is a cytochrome P-448 type enzyme as seen from its peak in the reduced-carbon monoxide difference spectrum, and this is also reflected in the pattern of benzo(a)pyrene metabolites produced which are more similar to those produced by a mammalian cytochrome P-448 type enzyme than by a mammalian cytochrome P-450 type (Wiseman & Woods 1979).

Microsomal yeast cytochrome P-450 is capable of using cumene hydroperoxide as an oxygen and electron source to catalyse benzo(a)pyrene hydroxylation in the absence of NADPH and molecular oxygen (Wiseman & Woods 1979). The requirement of the purified reconstituted system for NADPH and oxygen could also be replaced by hydrogen peroxide generated *in situ* from a glucose/glucose oxidase system as well as cumene hydroperoxide (King *et al.* 1984). Both of these artificial systems resulted in higher maximum velocities for the enzyme than observed for the NADPH-supported reaction, although lower affinities of the enzyme for benzo(a)-pyrene were seen as indicated by higher apparent K_m values (King *et al.* 1984) (see Table 5.1). With both the cumene hydroperoxide and hydrogen

Table 5.1 — Benzo[*a*]pyrene hydroxylase activity of highly purified cytochrome P-448 (from yeasts) in a reconstituted system

	K_m (μM)	V_{max} (pmol 3-hydroxybenzo[*a*]pyrene/ min per nmol P-448)
NADPH-supported	33	16.7
Cumene hudroperoxide	125	21.9
Hydrogen peroxide *in situ*	200	33.7

The reconstituted system comprised P-448 (1 nmol), NADPH–P-450 reductase (1 unit), dilauroylphosphatidylcholine (30 μg), all per ml of assay mixture. The NADPH-supported system contained an NADPH-generating system of 4 mM NADPH, 20 mM glucose 6-phosphate and 8 units/ml glucose 6-phosphate dehydrogenase. Alternatively cumene hydroperoxide was used at 0.53 M. Hydrogen peroxide was generated in *in situ* from a glucose system of 2 mM glucose and 2 units/ml glucose oxidase. Benzo[*a*]pyrene was used in the range 0–160 μM in a 15 min assay. V_{max} and K_m were determined from Lineweaver–Burk plots. Data from King *et al.* 1984.

peroxide-supported reactions, the 3–hydroxylation of benzo(a)pyrene is slowed after a few minutes and a fluorescent peak at 540 nm begins to show (not seen with trhe NADPH-supported reaction), owing to the formation of a mixture of benzo(a)pyrene metabolites, including phenols, diols, and quinones (King *et al.* 1984). This is similar to results obtained with cytochrome P-450 of phenobarbital-induced rat liver, which shows much larger amounts of benzo(a)pyrene quinones from the cumene hydroperoxide-supported reaction compared to the NADPH-supported case (Renneberg *et al.* 1981). The optimum temperature (37°C) for benzo(a)pyrene hydroxylation was the same for yeast cytochrome P-448-dependent benzo(a)pyrene hydroxylation supported by NADPH, cumene hydroperoxide, or hydrogen peroxide, although the hydrogen peroxide-supported reaction had a slightly higher pH optimum than that for the cumene hydroperoxide or NADPH-supported reactions which is at pH 6.5–7.

As expected for a cytochrome P-450 substrate, benzo(a)pyrene has an effect in causing a shift in the spin state of yeast cytochrome P-448 to a higher spin form. Purified cytochrome P-448 was isolated by King *et al.* (1984) as 94% low spin form at 22°C. On binding benzo(a)pyrene the temperature-dependent spin state equilibrium was shifted to 18% high spin at 22°C and 42% high spin at 34°C (from 16%). The modulation of the spin-state to high spin by substrate is thought to be important in the catalytic mechanism of cytochrome P-450 enzymes by controlling the redox potential of the cytochrome as discussed in section 5.1.2.

Yeast cytochrome P-448 has previously been shown to be modulated from low to high spin by the addition of the substrate lanosterol, although no quantitative measurement was attempted (Yoshida & Kumaoka 1975a, b).

We have atempted to measure the changes in redox potential of yeast cytochrome P-448 on benzo(a)pyrene-binding using the spectophotometric method of Sligar *et al.* (1979). A value of −305 mV was found for the free yeast enzyme (King 1982) which is close to the value of −300 mV determined for rat liver cytochrome P-450 (Sligar *et al.* 1979) and to the value of −303 mV determined for cytochrome $P\text{-}450_{cam}$ from *Pseudomonas putida* (Sligar 1976). However, a value for benzo(a)pyrene-bound cytochrome P-448 could not be obtained as the absorbance of benzo(a)pyrene itself interfered with the spectral determination of the fraction of cytochrome P-448 oxidized or reduced, making it impossible to determine the redox potential. The addition of triton X-100 to purified yeast cytochrome P-448 resulted in a shift in the mid-point redox potential of cytochrome P-448 to −334 mV (King 1982). The addition of triton X-100 results in stabilization of the yeast cytochrome P-448 (Azari & Wiseman 1980). This stabilization is thought to be partly due to triton X-100 binding causing a change in the spin-state to low spin which may protect the sulphydryl ligand of the haem iron. The modulation of the spin-state of yeast cytochrome P-448 from high to low spin by triton X-100 has also been shown by Tamura *et al.* (1981). Azari & Wiseman (1981) also reported that triton X-100 acts as a type II substrate in causing a slowed rate of reduction of cytochrome P-448, which is due to modulation of the enzyme to a lower spin state. The lower value of mid-point redox potential in the presence of triton X-100 is consistent with all of these observations.

The benzo(a)pyrene hydroxylase activity of the purified reconstituted cytochrome P-448 system can be inhibited by compounds which bind to cytochrome P-448 (King *et al.* 1984). Lansterol is a good inhibitor of benzo(a)pyrene hydroxylase as might be expected for another substrate of yeast cytochrome P-448. Dimethylnitrosamine also inhibits benzo(a)pyrene hydroxylase, although only at high concentrations. The specific inhibitor of rat liver cytochrome P-448 enzymes, 9-hydroxyellipticine (Delaforge *et al.* 1980) showed no inhibition of yeast benzo(a)pyrene hydroxylase (King *et al.* 1984).

Flavonoid compounds interact with benzo(a)pyrene hydroxylase from mammalian hepatic tissue as shown by Huang *et al.* (1981). Flavone stimulates the benzo(a)pyrene hydroxylation activity of the cytochrome P-450 isoenzymes LM3c and LM4, yet inhibits the same activity by the isoenzymes LM2, LM3b, or LM6. Also, 7,8-benzoflavone stimulates the benzo(a)pyrene hydroxylase activity of cytochrome P-450 LM3c, yet inhibits cytochrome P-450 LM6. The yeast enzyme is inhibited by both flavone and 7,8-benzoflavone in a non-competitive manner (K_i values of 600 μM and 176 μM respectively, King *et al.* 1984). In this respect yeast cytochrome P-448 differs greatly from cytochrome P-450 LM4, the major cytochrome P-448 form in rabbit liver, and appears to resemble cytochrome P-450 LM6 (another cytochrome P-448 form) which is induced by 2,3,7,8-tetrachlorodibenzo-*p*-dioxin (TCDD), particularly in neonatal rabbits (Norman *et al.* 1978).

The benzo(a)pyrene hydroxylase activity of yeast cytochrome P-448 has been used to study the effects of immobilization of this enzyme on a variety of supports (Azari & Wiseman 1982b). In this work, a purified reconstituted system of cytochrome P-448 with NADPH : cytochrome P-450 reductase and phospholipid was compared to microsomal fraction and permeabilized whole cells. Entrapment in calcium alginate beads was found to be especially useful, with the kinetics of hydroxylation similar to those of the free enzyme system with all three forms of the enzyme (purified, microsomal, and permeabilized cells). The purified reconstituted system was also successfully immobilized on cyanogen bromide-activated Sepharose 4B and entrapped in polyacrylamide gel. Again, good activity of the immobilized system was obtained. The immobilization in all forms of the enzyme allowed improved stability on storage at 4°C (Azari & Wiseman 1982b).

The activity of yeast cytochrome P-448 towards other xenobiotics has not been studied in as much detail as its activity towards benzo(a)pyrene. The hydroxylation of aniline and the demethylation of aminopyrine by yeast cytochrome P-448 was reported by Yoshida & Kumaoka (1975a,b). The demethylation of aminopyrine was also reported by Sauer *et al.* (1982) together with the demethylation of caffeine and *p*-nitroanisole. Most activity was seen with aminopyrine, less with *p*-nitroanisole, and the least with caffeine. The Nash assay for formaldehyde was used to measure the level of metabolism in these experiments. The *N*-demethylation of aminopyrine, benzphetamine, and dimethylnitrosoamine was sought using yeast microsomal fraction containing a high concentration of cytochrome P-448, and also using a purified reconstituted system by King (1982). This work again used the Nash assay in attempts to measure formaldehyde produced as a measure of enzyme activity. However, in these experiments no metabolism could be observed, and equivocal results were obtained with dimethylnitrosamine. The Nash assay technique is a relatively insensitive assay, and thus a small amount of metabolism my go undetected. King (1982) also attempted to measure aniline hydroxylation and the *O*-deethylation of ethoxyresorufin by yeast cytochrome P-448, both with negative results.

Wiseman *et al.* (1975) noted that yeast microsomal fraction demonstrated biphenyl hydroxylase activity thought to be due to a cytochrome P-450. However, this work could not be confirmed by Woods (1979) who used several different techniques in attempts to demonstrate biphenyl metabolism by yeast cytochrome P-448. Dodge *et al.* (1979) could detect only trace production of 2- and 4-hydroxybiphenyls after growing yeast in the presence of biphenyl. Smith *et al.* (1980) were also unable to detect biphenyl metabolites after growing *S. cerevisiae* in biphenyl-containing medium. King (1982) demonstrated biphenyl hydroxylase activity in intact yeast spheroplasts but could not detect any activity after lysis of the spheroplasts and subsequent cell fractionation. Similarly, a highly active biphenyl hydroxylase is present in the fungus *Cunninghamella elegans*, but on preparation of a microsomal fraction from this organism no biphenyl activity could be detected, although the same microsomal fraction contained a cytochrome P-450 and could hydroxylate naphthalene and benzo(a)-pyrene (Dodge *et al.* 1979). Karenlampi & Hynninen (1981a) demonstrated that cells of *S. cerevisiae* produced benzoic acid from biphenyl and that hydroxylated biphenyl intermediates could not be detected (equivocal results were obtained with the fluorimetric biphenyl hydroxylase assay used by those mentioned above). Karenlampi & Hynninen (1981a) suggested that biphenyl might be rapidly metabolized to benzoic acid, or that intermediates might decompose during sample preparation, in either case rendering the detection of hydroxylated biphenyls difficult.

Cytochrome P-450 enzymes from *S. cerevisiae* have been reported to metabolize many other promutagens to active mutagenic species (see Chapter 6). Callen & Philpot (1977) demonstrated that promutagens could be activated by *S. cerevisiae* when cells were grown on 2% (w/v) glucose medium such that cytochrome P-450 was present. Under these conditions the promutagens, aflatoxin B_1, dimethylnitrosamine, β-naphthylamine, ethyl carbamate, cyclophosphamide, and dimethylsulphoxide were activated to species active genetically in the same cells. The metabolism of all of these compounds except aflatoxin B_1 was high in cells containing spectrally detectable cytochrome P-450, but negligible in cells grown on galactose medium where no cytochrome P-450 could be detected. Aflatoxin B_1 was metabolized equally well under both sets of conditions and may be metabolized by a non-cytochrome P-450 route. The addition of cumene hydroperoxide to whole cells of *S. cerevisiae* had no effect on the stability of the cytochrome P-450 in the cells and resulted in an increase in the genetic activity of the promutagens due to increased metabolism (Callen *et al.* 1978). This effect may be due to cumene hydroperoxide acting as an artificial oxygen and electron donor to the enzyme, resulting in a faster rate of metabolism as has been shown for benzo(a)pyrene metabolism *in vitro* (King *et al.* 1984). See Chapter 6.

Yeast cytochrome P-450 was also shown to be involved in the metabolism of seven halogenated aliphatic hydrocarbons to active mutagenic species (methylene chloride, halothane, chloroform, carbon tetrachloride, trichloroethylene, tetrachloroethylene, and S-tetrachloroethane) (Callen *et al.* 1980). Spectral evidence was obtained which indicated metabolism of these compounds occurred *via* a cytochrome P-450 enzyme (Callen *et al.* 1980).

Kelly & Parry (1983) have developed methods for the assay of genetically active species in yeast using conditions of growth in high glucose such that the cytochrome P-450 level is high during the assay. This system is discussed in detail in Chapter 6 by Dr Diane Kelly.

The activation of benzo(a)pyrene, 15,16-dihydro-11-methylcyclopenta(a)phenanthren-17-one, 2-naphthylamine, and cyclophosphamide by yeast cytochrome P-450 were demonstrated in this system (Kelly & Parry 1983). Similar methodology has been used by Del Carratore *et al.* (1983) to show the activation of dimethylnitrosamine and styrene to genetically active species in yeast. In this work, exponentially growing cells with a high level of cytochrome P-450 showed good activation of these compounds, whilst stationary phase cells with no detectable cytochrome P-450 showed very poor metabolism.

5.2.3 The induction of cytochrome P-450 in *S. cerevisiae* (see general review on gene induction in yeast — Chapter 3)

In an early report, Wiseman & Lim (1975) noted that the addition of phenobarbital to yeast growing aerobically in low-glucose medium (conditions under which cytochrome P-450 cannot normally be detected spectrally) resulted in the induction of cytochrome P-450 (to levels similar to those found at the time in yeast grown at high glucose concentration). However, no effect on cytochrome P-450 levels in yeast grown at high glucose concentration was observed after the addition of phenobarbital. No attempt was made however, to measure any differences in the enzymic activity of the cytochrome P-450 in yeast grown under these two sets of conditions.

Karenlampi *et al.* (1982) demonstrated the induction of cytochrome P-450, in low glucose-containing media, by xenobiotic compounds. Both Clophen A_{60} (a polychlorinated biphenyl mixture) and lindane increased the levels of cytochrome P-450 present in *S. cerevisiae* grown in 0.5% (w/v) or 1% (w/v) glucose medium. These workers demonstrated also that these compounds were capable of inducing cytochrome P-450 in other yeast species.

The induction of a more specific benzo(a)pyrene hydroxylase after growth of *S. cerevisiae* in the presence of benzo(a)pyrene was first noted by Woods & Wiseman (1980). This finding was confirmed and extended by King *et al.* (1982). The addition of benzo(a)pyrene to yeast during growth in 20% (w/v) glucose medium results in only a small increase in cytochrome P-450 levels determined spectrally (up to 30%), but results in a dramatic improvement in the apparent kinetics of benzo(a)pyrene hydroxylation as indicated by a decreased apparent K_m and increased maximum velocity. This results in a much more efficient benzo(a)pyrene hydroxylase being produced, and this effect is directly dependent on the concentration of benzo(a)pyrene added to the growth medium. No induction of cytochrome

P-450 levels or benzo(a)pyrene hydroxylase could be seen in yeast grown in low-glucose medium, even in the presence of high concentrations (95 μM) of benzo(a)pyrene. This is unlike the finding for phenobarbital by Wiseman & Lim (1975) and for Clophen A_{60} and lindane by Karenlampi *et al.* (1982).

Other compounds were also capable of inducing a more specific benzo(a)pyrene hydroxylase (King *et al.* 1982). Dimethylnitrosamine and, to a lesser extent, phenobarbital both resulted in induction in a similar manner to benzo(a)pyrene. 3-Methylcholanthrene pretreatment again resulted in a large decrease in apparent K_m, although in this case no effect on maximum velocity was observed. β-Naphthoflavone, known to be a potent inducer of cytochrome P-448 enzymes in mammalian hepatic tissue, was unable to induce yeast benzo(a)pyrene hydroxylase. Pretreatment with lanosterol resulted in no change in spectrally detectable cytochrome P-450 yet resulted in an enzyme preparation with an increased K_m for benzo(a)pyrene, possibly owing to the induction of a less specific enzyme.

These results show that at least one new form of cytochrome P-450 can be produced in the presence of inducing agent, and this seems likely to occur with compensating non-production of the existing form of the enzyme. By analogy with mammalian systems this is probably due to the biosynthesis of a different form of cytochrome P-450 with a greater specificity for benzo(a)-pyrene, suggesting that more than one form of cytochrome P-450 can exist in *S. cerevisiae*. Results of a similar nature were later reported for *Aspergillus ochraceus*, with induction of microsomal cytochrome P-450, with improved kinetic parameters, by benzo(a)pyrene (Dutta *et al.*, Ghosh *et al.* 1983).

Induction with benzo(a)pyrene cannot be achieved in non-growing yeast (King 1982), and a time course of benzo(a)pyrene induction in growing yeast showed an initial increase in cytochrome P-450 level which later falls to control levels (King & Wiseman 1983). Change in the benzo(a)pyrene hydroxylase kinetic parameters is not instantaneous, and requires at least four hours to reach maximal levels, probably reflecting the time necessary for new enzyme synthesis to occur (see Table 5.2). The amount of NADPH-:cytochrome P-450 reductase, the other enzymic component of the system, is unchanged by benzo(a)pyrene treatment (King & Wiseman 1983). The initial increase in spectrally detectable cytochrome P-450 levels may be due to the production of the new (more efficient) enzyme form occurring at a faster rate than loss of the less-efficient enzyme, thereby accumulating more total enzyme. Another report of benzo(a)pyrene induction of benzo(a)pyrene hydroxylase in a fungal organism, is for *Neurospora crassa* which has been shown to produce a benzo(a)pyrene hydroxylase when grown in the presence of benzo(a)pyrene (Lin & Kapoor 1979).

Karenlampi and Hynninen (1981) grew *S. cerevisiae* in the presence of chrysene, 3-methylcholanthrene, and hexachlorophene. In some cases second derivatives of the reduced carbon monoxide difference spectra revealed a splitting of the cytochrome P-450 band in two which may have been due to the presence of two forms of cytochrome P-450. However, the spectrum of a mixture of forms of cytochrome P-450 from Arochlor 1254-

Table 5.2 — Time course of induction of cytochrome P-448 by 95 μM benzo(a)pyrene in yeast

Incubation time (h)	Treatment	Amount of Cytochrome P-448 (nmol/g wet wt. of yeast)	Benzo(a)pyrene hydroxylase activity		
			K_m (μM)	V_{max} (pmol/h per nmol of P-448)	Correlation coefficient
0	Control	4.67±0.16	111	167	0.996
	+Benzo(a)pyrene	4.44±0.21	108	222	0.993
2	Control	4.00 0.18	100	192	0.998
	+Benzo(a)pyrene	5.64±0.41	74	250	0.992
4	Control	3.51±0.38	109	233	0.964
	+Benzo(a)pyrene	5.01±0.48	46	455	0.992
8	Control	3.08±0.38	105	263	0.995
	+Benzo(a)pyrene	5.04±5.04	39	455	0.994
19	Control	5.44±0.45	111	217	0.991
	+Benzo(a)pyrene	6.91±0.27	33	417	0.973
24	Control	4.10±0.35	111	222	0.999
	+Benzo(a)pyrene	3.96±0.34	39	455	0.983

Values for cytochrome P-448 are means ± S.D. (n=8). Kinetic parameters of benzo(a)pyrene hydroxlyase, K_m (μM) *and* V_{max} (pmol of 3-hydroxybenzo(a)pyrene/h per nmol of cytochrome P-448) were determined from Lineweaver–Burk plots by using a simple linear regression (n=5). Data from King and Wiseman (1983).

induced rat liver did not split into multiple bands on taking second derivatives. King (1982) examined the second derivative spectra of yeast grown with and without benzo(a)pyrene, and in neither case was any splitting of the bands seen, although a small shift in peak wavelength appeared to take place.

5.2.4 Binding studies with promutagens (putative carcinogens)

The binding of benzo(a)pyrene to yeast cytochrome P-450 has been studied extensively (Woods & Wiseman 1980, Azari & Wiseman 1982a, b, King *et al.* 1984). Benzo(a)pyrene gives rise to a type I binding spectrum with yeast microsomal cytochrome P-488, indicative of substrate binding (Woods & Wiseman 1980). With both yeast and rat liver microsomal cytochrome P-450 an extra peak is observed at 360 nm in the benzo(a)pyrene binding spectrum (Estabrook *et al.* 1978). This extra peak is also observed in the type I binding spectrum observed with purified yeast cytochrome P-448, and with purified cytochrome P-450 from phenobarbital induced rat liver, although with these purified enzymes this peak is seen at 367 nm (King *et al.* 1984). The spectral dissociation constants (K_s values) for the purified yeast and rat liver enzymes were 50 μM (yeast) and 31 μM (rat liver), and for microsomal enzymes 18 μM (yeast) and 5 μM (rat liver) (Azari & Wiseman 1982b). The comparison of the absorbance of the benzo(a)pyrene–enzyme complex with the carbon monoxide binding spectrum allows a calculation of the proportion of the enzyme that binds benzo(a)pyrene (using an extinction coefficient of 57 $mMcm^{-1}$ for A415–500 nm of the benzo(a)pyrene–enzyme complex, Estabrook *et al.* 1978). Purified yeast cytochrome P-448 bound 100% to benzo(a)pyrene (i.e. each enzyme molecule binds one benzo(a)pyrene molecule), whereas only 53% of the purified phenobarbital-induced cytochrome P-450 from rat liver bound benzo(a)pyrene.

The binding of benzo(a)pyrene to yeast cytochrome P-448 has also been studied by equilibrium gel filtration with microsomal fraction (Woods & Wiseman 1980) and highly purified enzyme (Wiseman & Azari 1982). This technique allows an estimation of the number of binding sites for benzo(a)-pyrene. The values determined for yeast cytochrome P-448 in microsomal and purified forms were six and one respectively. As yeast cytochrome P-488 binds 100% to benzo(a)pyrene in both microsomal and purified forms, the number of binding sites are six for microsomal fraction and one for purified enzyme. The number of binding sites for rat liver enzyme were also six and one for microsomal and purified enzymes respectively (Wiseman & Azari 1982). One binding site, at the active site, might be expected as found in the purified enzyme. The extra five binding sites found for the microsomal enzymes are clearly due to high-affinity binding probably to the lipid present, as affinity is lowered in purified enzyme.

Lanosterol also gives rise to a type I binding spectrum with purified cytochrome P-448 from yeast, again with an extra peak at 367 nm (King *et al.* 1984). A K_s value of 80 μM was determined for lanosterol binding to purified cytochrome P-448. A type I binding spectrum was also shown for lanosterol by Aoyama & Yoshida (1978a, b).

Kelly *et al.* (1985) have looked at the binding of several promutagens to yeast cytochrome P-448. Benzo(a)pyrene and cyclophosphamide gave type I binding spectra with yeast microsomal cytochrome P-448, whereas benzidine and 2-naphthylamine gave type II binding spectra, indicating binding at the haem group. A binding spectrum with dimethylnitrosamine could not be obtained with microsomal cytochrome P-448, but when purified enzyme was used, a weak type I binding spectrum was observed, with a K_s value of 220 μM. Only strong binding spectra can be detected with yeast microsomal cytochrome P-448 preparations, owing to the turbidity of the microsomal suspension used.

Cyclophosphamide is activated by phenobarbital-induced cytochrome P-450 from mammalian tissues (Gurtoo *et al.* 1978) rather than cytochrome P-448 forms of the enzyme, therefore the binding spectrum of cyclophosphamide with yeast cytochrome P-448 is surprising. However, Kelly & Parry (1983) demonstrated that *S. cerevisiae* containing cytochrome P-448 could activate cyclophosphamide to genetically active species as efficiently as an Arochlor 1254-induced 'S9' mix (rat liver cytochrome P-450) added to the yeast system. The presence of a type I binding spectrum suggests that yeast cytochrome P-448 can metabolize cyclophosphamide as well as benzo(a)-pyrene, and demonstrates the unique substrate specificity of the yeast enzyme.

The type II binding seen with benzidine and 2-naphylamine is indicative of binding at the haem group of the enzyme. Benzidine is not activated *in vitro* by the *N*-hydroxylation route, and *N*-acetylation may occur (Martin *et al.* 1982). Ethanol-induced rat liver cytochrome P-450 gave a higher mutagenic activity for benzidine in bacteria than β-naphthoflavone or phenobarbital induction (Phillipson & Ioannides 1983). Yeast cytochrome P-448 is also induced by ethanol (see section 5.2.1), and this may indicate a similarity in the specificity of these two ethanol induced enzymes. 2-Naphthylamine is activated by the *N*-hydroxylation route (Radomski & Brill 1970), and seems to be activated in yeast by cytochrome P-448 (Callen & Philpot 1977, Kelly & Parry 1983).

From these studies it is seen that binding spectra offer a route to study the interaction of promutagens with cytochrome P-448 enzymes. The metabolism of promutagens is obviously of crucial importance in the interpretation of yeast genotoxicity studies, as is the ability of promutagens to induce particular isoenzymes of cytochrome P-448 in the yeast. (See Chapter 6).

5.3 TECHNIQUES FOR THE STUDY OF YEAST CYTOCHROME P-450 ENZYMES

5.3.1 Small-scale yeast growth for the production of cytochrome P-450

The conditions necessary to grow yeast for the optimal production of cytochrome P-450 have been discussed in detail (section 5.2.1). We have routinely used *S. cerevisiae* strain N.C.Y.C. No. 754 which produces a large amount of cytochrome P-450 when grown in high glucose medium. The yeast is maintained on slopes of Sabouraud-dextrose agar and grown up at

30°C in shake flasks containing 20% (w/v) glucose 2% (w/v) mycological peptone, 1% (w/v) yeast extract, and 0.5% (w/v) sodium chloride. The shake speed is important as it determines the dissolved oxygen concentration in the flask which is critical for cytochrome P-450 production (Blatiak *et al.* 1985a). We have found that a shake speed of 120 rev/min in an orbital shaker is optimal with 250 ml flasks containing 100 ml of medium, although it is probably necessary to optimize this for any system used.

5.3.2 Spectral assay of cytochrome P-450

Cytochrome P-450 levels can be monitored spectrally by the method of Omura & Sato (1964) using whole yeast cells or any cellular fraction prepared. For whole yeast cells a concentration of 0.1 g wet weight/ml in 0.1 M phosphate buffer pH 7.2 is used. The yeast suspension is divided into two cuvettes, and a few milligrams of sodium dithionite is added to each cuvette to reduce the enzyme. A baseline is drawn in the range 400–500 nm using a recording spectrophotometer. The sample cuvette is removed and carbon monoxide bubbled through it for 30 seconds at approximately 1 bubble per second. The difference spectrum is then recorded again between 400 and 500 nm. The concentration of cytochrome P-450 can then be calculated from the difference in absorbance between 450 and 490 nm with reference to the baseline, using an extinction coefficient of 91 mM^{-1} cm^{-1}.

It is of interest in this connection that an entirely different, solid phase radioimmunoassay, method has been reported to assay cytochrome P-450 in *Candida maltosa* (Kargel *et al.* 1984).

5.3.3 Preparation of yeast microsomal fraction

Yeast is harvested by centrifugation, washed in 0.1 M tris/HCl buffer, pH 7.2, and disrupted using a water-cooled Vibro-mill disruptor. The yeast was mixed with glass beads (1.00–1.05 mm diameters) and milled for a total of 6 minutes. The mill is stopped after one minute to top up with glass beads, and then milling is carried out in one minute bursts with one minute intervals for cooling, After milling, the glass beads are washed with 0.1 M tris/HCl, pH 7.2, and the washings centrifuged at 10 000 g for 10 minutes to remove unbroken cells, debris, and large cellular organelles. The supernatant is then spun at 100 000 g (or more) for one hour to obtain a microsomal pellet. The microsomal pellet is then resuspended in the desired buffer and re-centrifuged to obtain a washed microsomal pellet. This washed pellet is then resuspended for use using a hand-held Potter–Elvejhem homogeniser.

5.3.4 Assay of NADPH : cytochrome P-450 (c) reductase

NADPH : cytochrome P-450 reductase is conveniently assayed by its ability to reduce cytochrome c in a spectrophotometric assay (Yoshida *et al.* 1974a, b). To each of two cuvettes is added 1 ml of 0.1 mM cytochrome c, 0.1 ml enzyme sample, and 1.8 ml phosphate buffer, pH 7.6 (50 mM). A baseline is established using the two cuvettes at 550 nm, and then the reaction is initiated by the addition of 0.1 ml of 30 mM NADPH to the sample cuvette and 0.1 ml buffer to the reference. The rate of increase in absorbance at 550 nm is measured, and the level of NADPH: cytochrome c reductase is calculated using the extinction coefficient of 18.5 $mM^{-1} cm^{-1}$ for cytochrome c reduced. Results are expressed as units of activity where one unit converts 1 μmole of cytochrome c to its reduced form per minute.

5.3.5 The purification of yeast cytochrome P-448

Cytochrome P-448, cytochrome b_5, NADPH: cytochrome P-450 reductase, and NADH: cytochrome b_5 reductase can all be purified from the same preparation of yeast microsomes as shown in Fig. 5.5. (King *et al.* 1984).

For purification of microsomal enzymes, microsomes are resuspended in 0.1 M phosphate buffer, pH 7.2 containing 20% (v/v) glycerol, 1 mM EDTA and 0.1% (w/v) reduced glutathione to a concentration of approximately 30 mg protein per ml. The enzymes are solubilized by stirring under nitrogen for one hour (at 4°C) after the addition of 1% (w.v) sodium cholate. After this the mixture is centrifuged at 160 900 g for one hour and supernatant retained. The supernatant is then subjected to ammonium sulphate fractionation, with the fraction from 35–65% of saturation retained. Firstly ammonium sulphate is added to 35% saturation and the precipitate removed by centrifugation at 133 500 g for 20 minutes, and then sufficient ammonium sulphate is added to reach 65% of saturation. The precipitate is collected by centrifugation at 44 000 g for 20 minutes and resuspended in 20 mM phosphate buffer, pH 7.0 containing 20% (v/v) glycerol, 1 mM EDTA, 1 mM glutathione, and 0.5% (w/v) sodium cholate, and this is dialysed overnight against two changes of 30 volumes of 10 mM phosphate buffer, pH 7.0 containing 20% glycerol, 1 mM EDTA, 1mM glutathione, and 0.3% (w/v) sodium cholate. Insoluble material was removed by centrifugation (160 500 g for one hour), the sodium cholate concentration adjusted to 0.5% (w/v), and the sample applied to a column of 8-amino-n-octyl-Sepharose 4B previously equilibrated with 10 mM phosphate buffer containing 20% (v/v) glycerol, 1 mM EDTA, and 0.3% (w/v) sodium cholate. The 8-amino-*n*-octyl Sepharose 4B is prepared by coupling 1,8-diaminooctane to cyanogen bromide activated Sepharose 4B according to the manufacturer's instructions. Cytochrome P-448 is eluted (as a red band) using equilibration buffer also containing 0.1% (v/v) emulgen 911. Fractions containing cytochrome P-448 are then applied directly to hydroxylapatite column pre-equilibrated with 10 mM phosphate buffer, pH 7.0 containing 20% (v/v) glycerol. The column is washed with 30 mM phosphate buffer, pH 7.0 containing 20% (v/v) glycerol and 0.2% (v/v) emulgen 911. Cytochrome P-448 is then eluted with 100 mM phosphate buffer, pH 7.0 containing 20% (v/v) glycerol and 0.2% (v/v) emulgen 911. The cytochrome P-448-containing fractions are then pooled, diluted 10-fold with 20% (v/v) glycerol containing 0.2% (v/v) emulgen 911, and applied to a carboxymethyl-Sephadex C50 column pre-equilibrated with 10 mM phosphate buffer, pH 7.0 containing 20% (v/v) glycerol, 0.2% (v/v) emulgen 911. After washing with equilibration buffer and then 40 mM phosphate (20% glycerol, 0.2% emulgen 911) cytochrome P-448 was eluted with 100 mM phosphate buffer, pH 7.0 (20% glycerol, 0.2% emulgen 911). The cytochrome P-448 eluted was dialysed against two

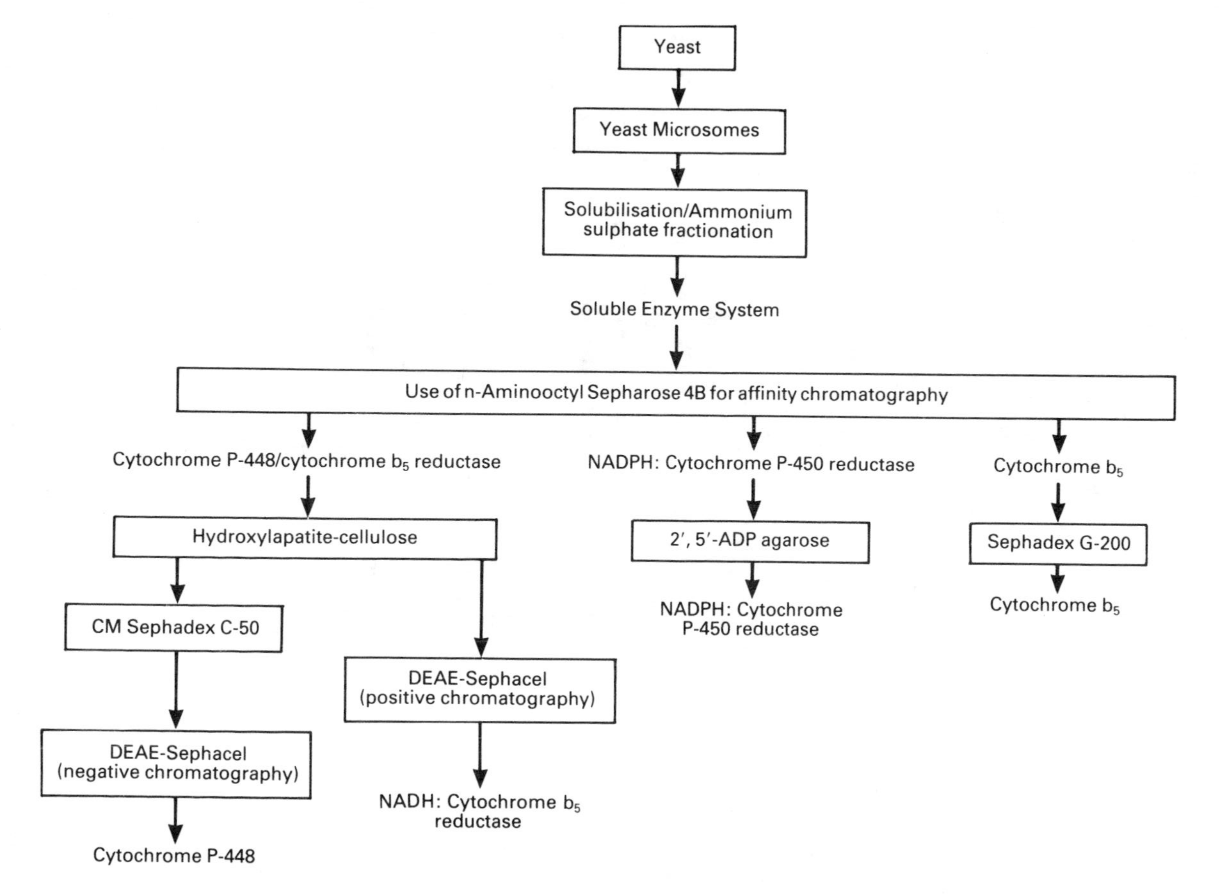

Fig. 5.5 — The scheme for the purification of cytochrome P-448, cytochrome b_5, NADPH–cytochrome P-450 reductase, and NADPH–cytochrome b_5 reductase from the same preparation of yeast microsomes isolated from *Saccharomyces cerevisiae*.

changes of 30 volumes of 5 mM phosphate buffer, pH 7.7 (20% glycerol, 0.2% emulgen 911), and then diluted (×2) with 20% glycerol before application to a DEAE-Sephacel column pre-equilibrated with the dialysis buffer. The majority of cytochrome P-448 does not bind, and is eluted directly from the column. Finally, the cytochrome P-448 is concentrated and non-ionic detergent removed by washing on a hydroxylapatite column before eluting with 100 mM phosphate buffer, pH 7.0. This procedure results in material of approximately 90% purity.

NADPH: cytochrome P-450 reductase can also be prepared using this procedure. After elution of cytochrome P-448 from the affinity column NADPH: cytochrome P-450 reductase can be eluted using the cytochrome P-448 elution buffer also containing 0.2% deoxycholate and 2 μM FMN. This material can then be purified further, using a 2′,5′-ADP-agarose column washed with 0.3 M phosphate buffer, pH 7.7 (20% glycerol, 0.1% emulgen 911), and the reductase can be eluted with this buffer containing 0.2 (w/v) deoxycholate and 5 mM 2′-AMP.

5.3.6 Fluorimetric assay for benzo(a)pyrene hydroxylase

Benzo(a)pyrene hydroxylase activity can be measured by the fluorimetric assay of Dehnen *et al.* (1973) as modified by Wiseman & Woods (1979). This method measures the formation of 3-hydroxybenzo(a)pyrene through its fluorimetric peak at 525 nm (467 nm excitation). The substrate, enzyme and cofactor solutions (Benzo(a)pyrene 20–160 μM added as a solution in dimethylformamide; NADPH, 4 mM) are incubated at 37°C and the reaction terminated by the addition of an equal volume of ice-cold acetone. Precipitated protein is removed by centrifugation and an aliquot (0.6 ml) of the supernatant is added to 1.4 ml of triethylamine (8.5% in water). This mixture is centrifuged again to remove any residual precipitate, a procedure which greatly reduces scatter in subsequent fluorescence measurements. Fluorescence is then measured at 467 nm excitation, 525 nm emission. Controls using boiled enzyme preparation, and complete reaction mixture stopped at zero time, should be run at each benzo(a)pyrene concentration used. Results are then calculated with reference to a 3-hydroxybenzo(a)-pyrene standard calibrated against quinine sulphate.

5.3.7 Binding spectra of cytochrome P-450 — compound complexes

Binding spectra are obtained by recording difference spectra between 350 and 500 nm after addition of the test compound to the enzyme preparation in the sample cuvette in increasing amounts. An equivalent volume of solvent is added to the enzyme preparation in the reference cuvette. A concentration of cytochrome P-450 of 1 μM is used. A split-cell technique is used to overcome interference from the absorbance of the test compounds themselves (Azari & Wiseman 1982b). In this technique the two compartment cells contain buffer in one compartment and enzyme preparation in the other. In the sample cuvette the test compound is added to the enzyme compartment, and in the reference cuvette the compound is added to the buffer compartment. This method has been described in detail by Azari & Wiseman (1982b).

5.4 CONCLUSIONS

The occurrence of cytochrome P-450 enzymes in the yeast *Saccharomyces cerevisiae* is now well established, and the conditions required for their biosynthesis and accumulation in yeast to high levels have been described. Fermentative growth is essential for the accumulation of large amounts of cytochrome P-450, although the reasons for this remain unclear. Oxygen, glucose, ethanol, and probably haem all play roles in regulating cytochrome P-450 production, as has been shown for other haemoproteins in *S. cerevisiae* (Hortner *et al.* 1982; Blatiak *et al.* 1987).

Yeast cytochrome P-450 enzymes are involved in the metabolism of at least one endogenous substrate, lanosterol, and are essential for ergosterol biosynthesis and the production of functional membranes. They are also capable of the metabolism of xenobiotics such as benzo(a)pyrene and the activation of many promutagens to genetically active species. This has led to the development of yeast test systems which do not require added mammalian enzymes for the activation of promutagens. However, it is apparent from the data discussed here that yeast cytochrome P-450 enzymes are not identical to those from mammalian sources and have different substrate specificities. One way of improving a yeast test system with a specified cytochrome P-450 activity in the cell might be to use a cloned mammalian enzyme expressed in the yeast test cell. In this respect it is notable that a 3-methylcholanthrene-induced cytochrome P-448 from rat liver has been cloned and expressed in *S. cerevisiae* (Oeda *et al.* 1985), and this enzyme is functional in drug metabolism (Sakaki *et al.* 1985, 1986).

REFERENCES

Alexander, K. T. W., Mitropoulos, K. A., & Gibbons, G. (1974) *Biochem. Biophys. Res. Commun.* **60** 460–467.

Ambike, S. H., Baxter, R. M., and Zahid, N. D. (1970) *Phytochemistry* **9** 1953–1958.

Aoyama, Y. & Yoshida, Y. (1978a) *Biochem. Biophys. Res. Commun.* **82** 33–38.

Aoyama, Y. & Yoshida, Y. (1978b) *Biochem. Biophys. Res. Commun.* **85** 28–34.

Aoyama, Y., Okikawa, T., and Yoshida, Y. (1981) *Biochim. Biophys, Acta* **666** 596–601.

Aoyama, Y., Yoshida, Y., Hata, S., Nishino, T., & Katsuki, H. (1983a) *Biochem. Biophys. Res. Commun.* **115** 642–647.

Aoyama, Y., Yoshida, Y., Hata, S., Nishino, T., Katsuki, H., Maitra, U. S., Mohan, V. P., & Sprinson, D. B. (1983b) *J. Biol. Chem.* **258** 9040–9042.

Aoyama, Y., Yoshida, Y., & Sato, R. (1984) *J. Biol. Chem.* **259** 1661–1666.

Appleby, C. A. (1978) *Meth. Enzymol.* **52** 157–166.
Asperger, O., Naumann, A., & Kleber, H. P. (1981) *FEMS Microbiol. Letters* **11** 309–312.
Auret, B. J., Boyd, D. R., Robinson, P. M., & Watson, C. G. (1971) *Chem. Commun.* **1971** 1585–1587.
Azari, M. R. & Wiseman, A. (1980) *Biochem. Soc. Trans.* **8** 713–714.
Azari, M. R. & Wiseman, A. (1981) *Anal. Biochem.* **117** 406–409.
Azari, M. R. & Wiseman, A. (1982a) *Anal. Biochem.* **122** 129–138.
Azari, M. R. & Wiseman, A. (1982b) *Enzyme Microb. Technol.* **4** 401–404.
Azari, M. R. & Wiseman, A. (1982c) *Biochem. Soc. Trans.* **10** 133–135.
Berg, A. (1982) *Biochem. Biophys. Res. Commun.* **105** 303–311.
Berg, A. & Rafter, J. J. (1981) *Biochem. J.* **196** 781–786.
Berg, A., Carlstom, K., Ingelman-Sundberg, M., Rafter, J. J., & Gustafsson, J. A. (1977) In: *Microsomes and drug oxidations* (Ullrich, V., ed.) pp. 377–384. Pergamon Press, Oxford.
Bergersen, F. J. & Turner, G. L. (1975) *J. Gen. Microbiol.* **91** 345–355.
Bertrand, J. C., Bazin, H., Zacek, M., Gilewicz, M., & Azoulay, E. (1979) *Eur. J. Biochem.* **93** 237–243.
Bertrand, J. C., Mattei, G., Parra, C., Giordani, R., & Gilewicz, M. (1984) *Biochimie* **66** 583–588.
Blatiak, A., Gondal, J. A., & Wiseman, A. (1980) *Biochem. Soc. Trans.* **8** 711–712.
Blatiak, A., King, D. J., Wiseman, A., Salihon, J., & Winkler, M. (1985a) *Enzyme Microbial Technol.* **7** 553–556.
Blatiak, A., King, D. J., Wiseman, A. (1985b) *Biochem. Soc. Trans.* **13** 924.
Blatiak, A., King, D. J., Wiseman, A., & Winkler, M. A. (1987) *J. Gen. Microbiol.* **133** 1053–1059.
Bosterling, B., Trudell, J. R., Trevor, A. J., & Bendix, M. (1982) *J. Biol. Chem.* **257** 4375–4380.
Bostrom, H. (1983) *J. Biol. Chem.* **258** 15091–15097.
Breskvar, K. (1983) *J. Steroid Biochem.* **18** 51–53.
Breskvar, K. & Hudnik-Plevnik, T. (1977) *Biochem. Biophys. Res. Commun.* **74** 1192–198.
Breskvar, K. & Hudnik-Plevnik, T. (1981) *J., Steroid Biochem.* **14** 395–399.
Bresnick, E. (1978) In: *Advances in enzyme regulation* (Weber, G. ed.) Vol. 16, pp. 347–361. Pergamon Press, New York.
Broadbent, D. A. & Cartwright, N. J. (1974) *Microbios* **9** 119–131.
Brunstrom, A. & Ingelmann-Sundberg, M. (1980) *Biochem. Biophys. Res. Commun.* **95** 431–439.
Burnet, F., Darby, N., & Lodola, A. (1986) *Bio Essays* **4** 231–232.
Callen, D. F. & Philpot, R. M. (1977) *Mutation Res.* **45** 309–324.
Callen, D. F., Wolf, C. R., & Philpot, R. M. (1978) *Biochem. Biophys. Res. Commun.* **83** 14–20.
Callen, D. F., Wolf, C. R., & Philpot, R. M. (1980) *Mutation Res.* **77** 55–63.
Capdevila, J., Estabrook, R. W., & Prough, R. A. (1980) *Arch. Biochem. Biophys.* **200** 186–195.
Cardini, G. & Jurtshuk, P. (1970) *J. Biol. Chem.* **245** 2789–2796.

Cartledge, T. G., Lloyd, D., Erecinska, M., & Chance, B. (1972) *Biochem. J.* **130** 739–747.

Cerniglia, C. E. & Gibson, D. T. (1978) *Arch. Biochem. Biophys.* **186** 121–127.

Cerniglia, C. E. & Gibson, D. T. (1980) *J. Biol. Chem.* **255** 5159–5163.

Cerniglia, C. E., Dodge, R. H., & Gibson, D. T. (1982) *Chem. Biol. Interact.* **38** 161–173.

Chiang, J. Y. L. (1981) *Arch. Biochem. Biophys.* **211** 662–673.

Ciejan, L., Beattie, D. S., Gollub, E. G., Liu, B. K. P., & Sprinson, D. B,. (1980) *J. Biol. Chem.* **255** 1312–1316.

Cinti, D. L., Silgar, S. G., Gibson, G. G., & Schenkman, J. B. (1979) *Biochem.* **18** 3642–3650.

Conney, A. H., Wood, A. W., Levin, W., Lu, A. Y. H., Chang, R. L., Wislocki, R. G., Goode, R. L., Holder, G. M., Dansette, P. M., Yagi, H., & Jerina, D. M. (1977) In: *Biological reactive intermediates* (Jollow, D. J., ed.) Plenum Press, New York.

Coon, M. J., Koop, D. R., Persson, A. V., & Morgan, E. T. (1980) In: *Biochemistry, biophysics and regulation of cytochrome P-450*, pp. 7–16 (Gustafsson, J. A., ed.) Elsevier, Amsterdam.

Coon, M. J., Koop, D. R., Reeve, L. E., & Crump, B. L. (1984) *Fundamental Appl. Toxicol.* **4** 134–143.

Coon, M. J., Black, S. D., Fujita, V. S., Koop, D. R., & Tarr, G. E. (1985) In: *Microsomes and drug oxidations* (ed. Boobis, A. R.) pp. 42–51, Taylor & Francis, London.

Correira, M. A. & Mannering, G. J. (1973) *Mol. Pharmacol.* **9** 455–469.

Coulson, C. J., King, D. J., & Wiseman, A. (1984) *Trends in Biochemical Sciences* **9** 446–449.

Danilov, V., Baranova, N., Ismailova, A., & Egorov, N. (1982) *Eur. J. Appl. Microbiol. Biotechnol.* **14** 125–129

Dehnen, W., Tomingas, R., & Roos, J. (1973) *Anal. Biochem.* **53** 373–383.

Delaforge, M., Ioannides, C., & Parke, D. V. (1980) *Chem. Biol. Interact.* **32** 101–110.

Del Carratore, R., Bronzetti, G., Bauer, C., Corsi, C., Nieri, R., Paolini, M., & Giagoni, P. (1983) *Mutation Res.* **121** 117–123.

Del Carratore, R., Morganti, C., Gali, A., & Bronzetti, G., (1984) *Biochem. Biophys. Res. Commun.* **123** 186–193.

De Pierre, J. W. & Ernster, L. (1980) In: *Microsomes, drug oxidations and chemical carcinogenesis* (Coon, M. J. ed.) pp. 431–443.

Desjardins, A. E., Matthews, D. E., & Van Etten, H. D. (1984) *Plant Physiol.* **75** 611–616.

Dodge, R. M., Cerniglia, C. E., & Gibson, D. T. (1979) *Biochem. J.* **178** 223–230.

Duppel, W., Lebeault, J. M., & Coon, M. J. (1973) *Eur. J. Biochem.* **36** 583–592.

Dus, K. (1975) *Adv. Exp. Biol. Med.* **58** 287–309.

Dus, K., Litchfield, W. J., Hippenmayer, P. J., Bumpus, J. A., Obidoa, O., Spitsberg, V., & Jefcoate, C. R,. (1980) *Eur. J. Biochem.* **111** 307–314.

Dutta, D., Ghosh, D. K., Mishra, A. K., & Samanta, T. B. (1983) *Biochem. Biophys. Res. Commun.* **115** 692–699.

Edelson, J. & McMullen, J. P. (1977) *Drug Met. Disp.* **5** 185–190.

Ellin, A. & Orrenius, S. (1975) *FEBS Letters* **50** 378–381.

Estabrook, R. W., Franklin, M. R., Cohen, B., Shigematsu, A., & Hildebrandt, A. G. (1971) *Metabolism* **20** 187–199.

Estabrook, R. W., Werringloer, J., Capdevila, J., & Prough, R. A. (1978) In: *Polycyclic hydrocarbons and cancer* (Gelboin, H., ed.) Vol. 1, pp. 285–319. Academic Press, New York.

Ferris, J. P., McDonald, L. H., Patrie, M. A., & Martin, M. A. (1976) *Arch. Biochem. Biophys.* **175** 443–452

Foldes, R. L., Hines, R. N., Ho, K. L., Shen, M. L., Nagel, K. B., & Bresnick, E. (1985) *Arch. Biochem. Biophys.* **239** 137–146.

French, J. S., Guengerich, F. P., & Coon, M. J. (1980) *J. Biol. Chem.* **255** 4112–4119.

Fujii-Kuriyama, Y., Mizukami, Y., Kawajiri, K., Sogawa, K., & Muramatsu, M. (1982) *Proc. Natl. Acad. Sci.* **79** 2793–2797.

Fujii-Kuriyama, Y., Sogawa, K., Suwa, Y., Kawajiri, K., & Gotoh, O. (1984) In: *Proceedings of IUPHAR 9th International Congress of Pharmacology, London*, 1984, Vol. 3, pp. 211–217 (Paton, W. ed.) Macmillan, London.

Gallo, M., Bertrand, J. C., & Azoulay, E. (1971) *FEBS Letters* **19** 45–49.

Ghosh, D. K., Dutta, D., Samanta, T. B., & Mishra, A. K. (1983) *Biochem. Biophys. Res. Commun.* **113** 497–505.

Gibbons, G. F., Pullinger, C. R., & Mitropoulos, K. A. (1979) *Biochem. J.* **183** 309–315.

Gibson, G. G. (1985) In: *Microsomes and drug oxidations* (Boobis, A. R. ed.) pp. 33–41, Taylor & Francis, London.

Gibson, G. G., Cinti, D. L., Sligar, S. G., & Schenkman, J. B. (1980) *J. Biol. Chem.* **255** 1867–1873.

Gibson, M., Soper, C. J., Parfitt, R. T., & Sawell, G. J. (1984) *Enzyme Microb. Technol.* **6**, 471–475.

Gilewicz, M., Zacek, M., Bertrand, J. C., & Azoulay, E. (1979) *Can. J. Microbiol.* **25** 201–206.

Gillette, J. R. (1966) *Adv. Pharmacol.* **4** 219–261.

Golbeck, J. H. & Cox, J. C. (1984) *Biotechnol. Bioeng.* **26** 434–441.

Gonzalez, F. J. & Nerbert, D. W. (1985) *Nucleic Acids Res.* **13** 7269–7288.

Gonzalez, F. J., Tukey, R. H., & Nerbert, D. W. (1984) *Mol. Pharmacol.* **26** 117–121.

Guarente, L. & Mason, T. (1983) *Cell* **32** 1279–1286.

Gudenus, R., Spence, A., Hartig, A., Smith, M., & Ruis, H. (1984) *Curr. Genet.* **8** 45–48.

Gunsalus, I. C., Pederson, T. C., & Sligar, S. G. (1975) *Ann. Rev. Biochem.* **44** 377–401

Gurtoo, H. L., Dahms, R., Hipkens, J., & Vaught, J. B. (1978) *Life Sci.* **22** 45–52.

Gustafsson, J. A., Rondahl, L., & Bergman, J. (1979) *Biochem.* **18** 865–870.

Gustafsson, J. A., Berg, A., & Rafter, J. (1980) In: *Microsomes, drug oxidations and chemical carcinogenesis* (Coon, M. J. ed.) Vol. 1, pp. 15–26. Academic Press, New York.

Haniu, M., Ryan, D. E., Levin, W., & Shiveley, J. E. (1984) *Proc. Natl. Acad. Sci.* **81** 4298–4301.

Harada, N. & Omura, T. (1983) *J. Biochem.* **93** 1361–1373.

Hata, S., Nishino, T., Komori, M., & Katsuki, H. (1981) *Biochem. Biophys. Res. Commun.* **103** 272–277.

Hata, S., Nishino, T., Katsuki, H., Aoyama, Y., & Yoshida, Y. (1983) *Biochem. Biophys. Res. Commun.* **116** 162–166.

Hauraud, M. & Ullrich, V. (1985) *J. Biol. Chem.* **260** 15059–15067.

Hildebrandt, A. & Estabrook, R. W. (1971) *Arch. Biochem. Biophys.* **143** 66–79.

Holloway, P. W. (1971) *Biochem.* **10** 1556–1560.

Hortner, H., Ammerer, G., Hartter, E., Hamilton, B., Rytka, J., Bilinski, T., & Ruis, H. (1982) *Eur. J. Biochem.* **128** 179–184.

Hrycay, E. G. & O'Brien, P. J. (1972) *Arch. Biochem. Biophys.* **153** 480–488.

Huang, M. T., Chang, R. L., Fortner, J. G., & Conney, A. H. (1981) *J. Biol. Chem.* **256** 6829–6836.

Imai, Y. (1979) *J. Biochem.* **86** 1697–1707.

Ingelman-Sundberg, M. & Hagbjork, A. (1982) *Xenobiotica* **12** 673–686.

Ishidate, K., Kawaguchi, K., Tagawa, K. & Hagihara, B. (1969a) *J. Biochem.* **65** 375–383.

Ishidate, K., Kawaguchi, K., & Tagawa, K. (1969b) *J. Biochem.* **65** 385–392.

Iversen, P. L., Hines, R. N., & Bresnick, E. (1986) *Bio Essays* **4** 15–19.

Jerina, D. M. & Daly, J. W. (1974) *Science* **185** 573–582.

Kadlabur, F. F., Morton, K. C., & Ziegler, D. M. (1973) *Biochem. Biophys. Res. Commun.* **54** 1255–1261.

Kappeli, O. (1986) *Microbiological Reviews* **50** 244–258.

Karenlampi, S. O. & Hynninen, P. H. (1981a) *Chemosphere* **10** 391–396.

Karenlampi, S. O. & Hynninen, P. H. (1981b) *Biochem. Biophys. Res. Commun.* **100** 297–304.

Karenlampi, S. O., Marin, E., & Hanninen, O. O. P. (1980) *J. Gen. Microbiol.* **120** 529–533.

Karenlampi, S. O., Marin, E., & Hanninen, O. O. P. (1981) *Biochem. J.* **194** 407–413.

Karenlampi, S. O., Marin, E., & Hanninen, O. O. P. (1982) *Arch. Environ. Contam. Toxicol.* **11** 693–698.

Karenlampi, S. O., Nikkila, H., & Hynninen, P. J. (1986) *Biotechnol. Appl. Biochem.* **8** 60–68.

Kargel, E., Schmidt, M. E., Schunck, W. H., Riege, P., Manersberger, S., & Muller, H. G. (1984) *Analytical Letters* **17** 2011–2024.

Kawajiri, K., Yonekawa, H., Gotoh, O., Watanabe, J., Igarashi, S., & Tagashira, Y. (1983) *Cancer Research* **43** 819–823.

Kelly, D. E. & Parry, J. M. (1983) *Mutation Res.* **108** 147–159.

Kelly, S. L., Kelly, D. E., King, D. J., & Wiseman, A. (1985) *Carcinogenesis* **6** 1321–1325.

King, D. J. (1982) PhD thesis, University of Surrey.

King, D. J. & Wiseman, A. (1983) *Biochem. Soc. Trans.* **11** 708–709.

King, D. J., Azari, M. R., & Wiseman, A. (1982) *Biochem. Biophys. Res. Commun.* **105** 1115–1121.

King, D. J., Wiseman, A., & Wilkie, D. (1983a) *Mol. Gen. Genet.* **192** 466–470.

King, D. J., Blatiak, A., & Wiseman, A. (1983b) *Biochem. Soc. Trans.* **11** 710.

King, D. J., Azari, M. R., & Wiseman, A. (1984) *Xenobiotica* **14** 187–206.

King, D. J., Wiseman, A., Kelly, D. E., & Kelly, S. L. (1985) *Curr. Genet.* **10** 261–267.

Kodama, O., Koichi, T., Akatsuka, T., & Vesugi, Y. (1982) *J. Pest. Sci.* **7** 517–521.

Kubota, S., Yoshida, Y., & Kumaoka, H. (1977) *J. Biochem.* **81** 187–195.

Kumaki, K., Sato, R., Kon, H., & Nerbert, D. W. (1978) *J. Biol. Chem.* **253** 1048–1058.

Kumar, A. & Padmanaban, G. (1980) *J. Biol. Chem.* **255** 522–525.

Labbe-Bois, R. & Volland, C. (1977) *Arch. Biochem. Biophys.* **179** 565–577.

Lavel, B. & Omenn, G. S. (1986) *Nature* **324** 29–34.

Laz, T. M., Pietras, D. F., & Sherman, F. (1984) *Proc. Natl. Acad. Sci.* **81** 4475–4497.

Lees, N. D., Kemple, M. D., Barbuch, R. J., Smith, M. A., & Bard, M. (1984) *Biochim. Biophys. Acta* **776** 105–112.

Levin, W., Wood, A. W., Lu, A. Y. H., Ryan, D., West, S., Conney, A. H., Thakker, D. R., Yagi, H., & Jerina, D. M. (1977) In: *ACS Symposium, Drug Metabolism Concepts* (Jerina, D. M. ed.) pp. 99–125. ACS, Washington.,

Levin, W., Thomas, P. E., Riek, L. M., Ryan, D. E., Bandiera, S., Haniu, M., & Shirley, J. E. (1985) In: *Microsomes and drug oxidations* (Boobis, A. R. ed.) pp. 13–22. Taylor & Francis, London.

Libor, S., Bloxsidge, J. P., Elvidge, J. A., Jones, J. R., Woods, L. F. J., & Wiseman, A. (1980) *Biochem. Soc. Trans.* **8** 99–100.

Lichtenberger, F. & Ullrich, V. (1977) In: *Microsomes and drug oxidations* (Ullrich, V. ed.) pp. 218–224. Pergamon Press, New York.

Lim, T.-K. (1976) PhD thesis, University of Surrey.

Lin, W. S. & Kapoor, M. (1979) *Current Microbiol.* **3** 177–180.

Lindenmayer, A. & Smith, L. (1964) *Biochim. Biophys. Acta* **93** 445–461.

Lu, A. Y. H. & Coon, M. J. (1968) *J. Biol. Chem.* **243** 1331–1342.

Lu, A. Y. H., West, S. B., Vore, M., Ryan, D., & Levin, W. (1974) *J. Biol. Chem.* **249** 6701–6709.

Mansuy, D., Cartier, M., Bertrand, J. C., & Azoulay, E. (1980) *Eur. J. Biochem.* **109** 103–108.
Marchal, R., Metche, M. & Vandecasteele, J. P. (1982) *J. Gen. Microbiol.* **128** 1125–1134.
Martin, C. N., Beland, F. A., Roth, R. N., & Kadlubar, F. (1982) *Cancer Res.* **42** 2678–2686.
Matson, R. S. Hare, R. S., & Fulco, A. J. (1977) *Biochim. Biophys. Acta* **487** 487–494.
Matthews, D. E. & Van Etten, H. D. (1983) *Arch. Biochem. Biophys,* **224** 494–505.
Mauersberger, S. & Matyashova, R. N. (1980) *Microbiol. (USSR)* **49** 498–503.
Mauersberger, S., Matyashova, R. N., Muller, H. G., & Losinov, A. B. (1980) *Eur. J. Appl. Microbiol. Biotechnol.* **9** 285–294.
Mauersberger, S., Schunck, W. H., & Muller, H. G. (1981) *Zeitschrift Allgemeine Microbiol.* **21** 313–321.
Mauersberger, S., Schunck, W. H., & Muller, H. G. (1984) *Appl. Microbiol. Biotechnol.* **19** 29–35.
MacDonald, T. L., Burka, L. T., Wright, S. T., & Guengerich, F. P. (1982) *Biochem. Biophys. Res. Commun.* **104** 620–625.
McIntosh, P. R., Kawato, S., Freedman, R. B., & Cheery, R. J. (1980) *FEBS Letters* **122** 54–58.
Misselwitz, R., Rein, H., Ristau, O., Janig, G. R., Zirwer, D., & Ruckpaul, K. (1980) *Studia biophysica* **79** 165–166.
Morita, T. & Mifuchi, I. (1984) *Chem. Pharm. Bulletin* **32** 1624–1627.
Muller, H. G., Schunck, W. H., Riege, P., & Honeck, H. (1979) *Acta Biol. Med. Germ.* **38** 345–349.
Muller, H. G., Schunck, W. H., Riege, P., & Honeck, H. (1982) In: *Cytochrome P-450, biochemistry, biophysics and environmental implications* (Hietanen, E. ed.) pp. 445–448. Elsevier, Amsterdam.
Muller, H. G., Schmidt, W. E., & Stier, A. (1985) *FEBS Letters* **187** 21–24.
Murphy, G., Vogel, G., Krippahl, G., & Lynen, F. (1974) *Eur. J. Biochem.* **49** 443–455.
Narhi, L. O. & Fulco, A. J. (1982) *J. Biol. Chem.* **257** 2147–2150.
Nebert, D. W. (1979) *Mol. Cell Biochem.* **27** 27–46.
Nebert, D. W. & Gonzalez, F. J. (1985) *Trends Pharmacol. Sci.* **6** 160–164.
Nebert, D. W., Bigelow, S. W., Okey, A. B., Yahagi, T., Mori, Y., Nagao, M., & Sugimura, T. (1979) *Proc. Natl. Acad. Sci.* **76** 5929–5936.
Nebert, D. W., Eisen, H. J., & Hankinsen, O. (1984) *Biochem. Pharmacol.* **33** 917–924.
Nelson, S. D. (1982) *J. Med. Chem.* **25** 753–765.
Nordblom, G. D., White, R. E., & Coon, M. J. (1976) *Arch. Biochem. Biophys.* **175** 524–533.
Norman, R. L., Johnson, E. F., & Muller-Eberhard, U. (1978) *J. Biol. Chem.* **253** 8640–8647.
Noshiro, M., Harada, N., & Omura, T. (1980) *J. Biochem.* **88** 1521–1525.

Oeda, K., Sakaki, T., & Ohkawa, H. (1985) *DNA* **4** 203–210.
O'Keefe, D. H., Ebel, R. E., & Peterson, J. A. (1978) *Meth. Enzymol.* **52** 151–157.
Omura, T. & Sato, R. (1964) *J. Biol. Chem.* **239** 2370–2379.
Ozols, J. (1986) *J. Biol. Chem.* **261** 3965–3979.
Parke, D. V. & Ioannides, C. (1982) *Adv. Exp. Med. Biol.* **136A** 23–38.
Parke, D. V., Ioannides, C., Iwasaki, K., & Lewis, D. F. V. (1985) In: *Microsomes and drug oxidations* (Boobis, A. R. ed.) pp. 402–413. Taylor & Francis, London.
Peterson, J., Ebel, R. E., O'Keefe, D. A., Matsubara, T., & Estabrook, R. W. (1976) *J. Biol. Chem.* **251** 4010–4016.
Phillips, I. R., Shephard, E. A., Rabin, B. R., Ashworth, A., & Pike, S. F. (1985) In: *Microsomes and drug oxidations* (Boobis, A. R. ed.) pp. 118–127. Taylor & Francis, London.
Phillipson, C. & Ioannides, C. (1983) *Mutation Res.* **124** 325–330.
Phillipson, C. E., Godden, P. M. M., Lum, P. Y., Ioannides, C., & Parke, D. V. (1984) *Biochem. J.* **221** 81–87.
Pierce, A. M., Mueller, R. B., Unrau, A. M., & Oehlschlager, A. C. (1978) *Can. J. Biochem.* **56**, 794–800.
Poole, R. K., Lloyd, D., & Chance, B. (1974) *Biochem. J.* **??** 201–210.
Poulos, T. L., Finzel, B. C., Gunsalus, I. C., Wagner, G. C., & Krant, J. (1985) *J. Biol. Chem.* **260** 16122–16130.
Powis, G. & Jansson, I. (1979) *Pharmacol. Therap.* **7** 297–311.
Prognon, P., Blais, J., Vigny, P., Averbeck, D., Averbeck, S., & Gond, A. (1984) *Il Farmaco Edizione Scientifica* **39** 739–751.
Qureshi, I., Lim, T. K., & Wiseman, A. (1980) *Biochem. Soc. Trans.* **8** 573–574.
Radomski, J. L. & Brill. E. (1970) *Science* **167** 992–993.
Rahimtula, A. D. & O'Brien, P. J. (1974) *Biochem. Biophys. Res. Commun.* **60** 440–447.
Rahimtula, A. D., O'Brien, P. J., Seifried, M. E., & Jerina, D. M. (1978) *Eur. J. Biochem.* **89** 133–141.
Renneberg, R., Capdevila, J., Chacos, N., Estabrook R. W., & Prough, R. A. (1981) *Biochem. Pharmacol.* **30** 843–848.
Reuttinger, R. T. & Fulco, A. J. (1981) *J. Biol. Chem.* **256** 5728–5734.
Reuttinger, R. T., Kim, B. H., & Fulco, A. J. (1984) *Biochim. Biophys. Acta* **801** 372–380.
Ristau, O., Rein, H., Greschner, S., Janig, G. R., & Ruckpaul, K. (1979) *Acta Biol. Med. Germ.* **38** 177–189.
Rogers, P. J. & Stewart, P. R. (1973) *J. Bacteriol.* **115** 88–97.
Ross, E. & Schatz, G. (1976) *J. Biol. Chem.* **251** 1997–2004.
Ryan, D. E., Thomas, P. E., & Levin, W. (1982) *Xenobiotica* **12** 727–744.
Sakaki, T., Oeda, K., Hiyoshi, M., & Ohkawa, H. (1985) *J. Biochem.* **98** 167–175.
Sakaki, T., Oeda, K., Yabusaki, Y., & Ohkawa, H. (1986) *J. Biochem.* **99** 741–749.

Salihon, J., Winkler, M. A., & Wiseman, A. (1983) *Biochem. Soc. Trans.* **11** 401.
Salihon, J., Winkler, M. A., & Wiseman, A. (1985) *Biochem. Soc. Trans.* **13** 526.
Sanglard, D., Kappeli, O., & Fiechter, A. (1984) *J. Bacteriol.* **157** 297–302.
Sato, R. & Omura, T. (1978) *Cytochrome P*-450, Kodansha—Academic Press.
Sauer, M., Kappeli, O., & Fiechter, A. (1982) In: *Cytochrome P-450, biochemistry, biophysics and environmental implications* (Hietanen, E. ed.) pp. 453–457. Elsevier, Amsterdam.
Schenkman, J. B., Remmer, H., & Estabrook, R. W. (1967) *Mol. Pharmacol.* **3** 113–123.
Schenkman, J. B., Sligar, S. G., & Cinti, D. L. (1981) *Pharmacol. Therap.* **12** 43–71.
Schunck, W. H., Riege, P., & Kuhl, R. (1978) *Pharmazie* **33** 412–414.
Schwartz, D., Pirrwitz, J., & Ruckpaul, K. (1982) *Arch. Biochem. Biophys.* **216** 322–328.
Sesardic, D., Boobis, A. R., McQuade, J., Baker, S., Lock, E. A., Elcombe, C. R., Robson, R. T., Hayward, C. & Davies, D. S. (1986) *Biochem. J.* **236** 569–577.
Shephard, E., Phillips, I. R., Pike, S. F., Ashworth, A., & Rabin, B. R. (1982) *FEBS Letters* **150** 375–380.
Shimakata, T., Mihara, K., & Sato, R. (1972) *J. Biochem.* **72** 1163–1174.
Shoun, H., Sudo, Y., Seto, Y., & Beppu, T. (1983) *J. Biochem.* **94** 1219–1229.
Sims, P., Grover, P. L., Swasiland, A., Pal, K., & Hewer, A. (1974) *Nature* **252** 326–328.
Silgar, S. G. (1976) *Biochem.* **15** 5399–5406.
Silgar, S. G., Cinti, D. L., Gibson, G. G., & Schenkman, J. B. (1979) *Biochem. Biophys. Res. Commun.* **90** 925–932.
Smith, R. V., Davis, P. J., Clark, A. M., & Glover-Milton, S. (1980) *J. Appl. Bacteriol.* **49** 65–73.
Sogawa, K., Gotoh, O., Kawajiri, K., & Fujii-Kuriyama, Y. (1984) *Proc. Natl. Acad. Sci.* **81** 5066–5070.
Stasiecki, P., Oesch, F., Bruder, G., Jarasch, E. D., & Franke, W. W. (1980) *Eur. J. Cell. Biol.* **21** 79–92.
Sugiyama, T., Miki, N. & Yamano, T. (1981) *J. Biochem.* **87** 1457–1467.
Suhura, K., Gomi, T., Sato, H., Itagaki, E., Takemori, S., & Katagiri, M. (1978) *Arch. Biochem. Biophys.* **190** 290–299.
Sunairi, M., Wantanabe, K., Takagi, M., & Yano, K. (1984) *J. Bacteriol.* **160** 1037–1040.
Takemori, S., Suhara, K., & Katagiri, M. (1978) In: *Cytochrome P*-450 (Sato, R. & Onura, T., eds.) pp. 164–184. Kodansha–Academic Press.
Takikawa, O., Gomi, T., Suhara, K., Hagaki, E., Takemori, S., & Katagiri, M. (1978) *Arch. Biochem. Biophys.* **190** 300–306.
Tamburini, P. P. & Gibson, G. G. (1983) *J. Biol. Chem.* **258** 13444–13453.

Tamaura, Y., Yoshida, Y., Sato, R., & Kumaoka, H. (1976) *Arch. Biochem. Biophys.* **175** 284–294.
Thomas, P. G., Haslam, J. M., & Baldwin, B. C. (1983) *Biochem. Soc. Trans.* **11** 713.
Thompson, E. A. & Siteri, P. K. (1974) *J. Biol. Chem.* **249** 5373–5378.
Tierney, B., Munzer, S., & Bresnick, E. (1983) *Arch. Biochem. Biophys.* **222** 826–835.
Trinn, M., Kappeli, O., & Fiechter, A. (1982) *Eur. J. Appl. Microbiol. Biotechnol.* **15** 64–68.
Trocha, P. J., Jasne, S. J., & Sprinson, D. B. (1977) *Biochemistry* **16** 4721–4726.
Van den Bossche, H. & Willemsens, G. (1982) *Arch. Int. Physiol. Biochim.* **90** B218–B219.
Van den Bossche, H., Willemsens, G., Cools, W., Cornelissen, F., Lauwers, W. F., & Van Cutsem, J. H. (1980) *Antimicrob. Agents Chemother.* **17** 922–928.
Van den Bossche, H., Willemsens, G., Cools, W., Marichal, P., Lauwers, W. F. (1983) *Biochem. Soc. Trans.* **11** 665–667.
Van den Bossche, H., Lauwers, W., Willemsens, G., Marichal, P., Cornelissen, F., & Cools, W. (1984) *Pestic. Sci.* **15** 188–198.
Vermillion, J. L. & Coon. M. J. (1978) *J. Biol. Chem.* **253** 8812–8819.
Wacket, L. P. & Gibson, D. T. (1982) *Biochem. J.* **205** 117–122.
White, R. E. & Coon, M. J. (1980) *Ann. Rev. Biochem.* **49** 315–356.
White, P. C., New, M. I., & Dupont, B. (1984) *Proc. Natl. Acad. Sci.* **81** 1986–1990.
Wiggins, T. E. & Baldwin, B. C. (1984) *Pestic. Sci.* **15** 206–209.
Wiseman, A. & Lim, T. K. (1975) *Biochem. Soc. Trans.* **3** 974–977.
Wiseman, A. & Woods, L. F. J. (1979) *J. Chem. Tech. Biotechnol.* **29** 320–324.
Wiseman, A. & Azari, M. R. (1982) *Biochem. Soc. Trans.* **10** 135–136.
Wiseman, A., Lim, T. K., & McCloud, C. (1975) *Biochem. Soc. Trans.* **3** 276–278.
Wiseman, A., McCloud, C., & Lim, T. K. (1976) *Biochem. Soc. Trans.* **4** 685–688.
Wiseman, A., Lim, T. K., & Woods, L. F. J. (1975) *Biochim. Biophys. Acta* **544** 615–623.
Woods, L. F. J. (1979) PhD thesis, University of Surrey.
Woods, L. F. J. & Wiseman, A. (1980) *Biochim. Biophys. Acta* **613** 52–61.
Wyman, J. F., Gollan, J. L., Settle, W., Farrell, G. C., & Correia, M. A. (1986) *Biochem. J.* **238** 837–846.
Yang, C. S. (1975) *FEBS Letters* **54** 61–64.
Yoshida, Y. & Kumaoka, H. (1972) *J. Biochem.* **71** 915–918.
Yoshida, Y. & Kumaoka, H. (1975a) *J. Biochem.* **78** 455–468.
Yoshida, Y. & Kumaoka, H. (1975b) *J. Biochem.* **78** 785–794.
Yoshida, Y. & Aoyama, Y. (1984) *J. Biol. Chem.* **259** 1655–1660.
Yoshida, Y., Kumaoka, H., & Sato, R. (1974a) *J. Biochem.* **75** 1201–1210.
Yoshida, Y., Kumaoka, H., & Sato, R. (1974b) *J. Biochem.* **75** 1211–1219.

Yoshida, Y., Aoyama, Y., Kumaoka, H., & Kubota, S. (1977) *Biochem. Biophys. Res. Commun.* **78** 1005–1010.

Yoshida, Y., Aoyama, Y., Nishino, Y., Katsuki, H., Maitra, U. S., Mohan, V. P., & Sprinson, D. B. (1985) *Biochem. Biophys. Res. Commun.* **127** 623–628.

Yoshioka, H., Morohashi, K., Sogawa, K., Yamane, M., Kominami, S., Takemori, S., Okada, Y., Omura, T., & Fujii-Kuriyama, Y. (1986) *J. Biol. Chem.* **261** 4106–4109.

Yu, C. A., Gunsalus, I. C., Katagiri, M., Suhara, K., & Takemori, S. (1974) *J. Biol. Chem.* **249** 94–107.

6

Detection of mutagens in yeast

Dr. Diane E. Kelly
Wolfson Institute of Biotechnology, The University , Sheffield

6.1 INTRODUCTION

6.1.1 Principles

Heritable alteration to DNA is more usually known as mutation, and in its widest sense this includes changes in the number of chromosomes, changes in gross structure of the chromosomes, and changes in the gene itself (Freese 1963).

The last three decades have seen much attention focused on identifying the wide range of agents in the environment which have the potential to induce heritable change in man.

Exposure to X-rays (Muller 1927) and nitrogen mustard (Auerbach 1947) could induce mutation in *Drosophilia* was known prior to the elucidation of the structure of DNA (Watson & Crick 1953), but with the understanding of the molecular basis of heredity came the realization that the molecular basis of mutation was a change in base sequence resulting in a change in the genetic code. Such a change in base sequence could involve either the substitution of one base for another, or the addition or deletion of bases.

Further stimulus for identifying the agents which are mutagenic and their mode of action came from the growing weight of evidence that supported the theory of a mutational origin to carcinogenesis.

In 1964 Brookes & Lawley demonstrated the binding of a polycyclic aromatic hydrocarbon to DNA. Before this many agents that were known to be carcinogenic had not been shown to interact with DNA. However, it was apparent that not all compounds were direct carcinogens. Sasaki & Yoshida (1935) had demonstrated that azo-dyes did not in general act at the site of contact of compounds with the organism, but rather at remote areas like the liver. The nature of indirect-chemical carcinogenesis has been the subject of considerable investigation; for example, in 1971 Malling, using a liver preparation in combination with a strain of *Neurospora crassa*, demonstrated that the carcinogen dimethylnitrosamine required metabolism to an ultimate mutagenic species.

This and other investigations have shown the importance of considering the fate of xenobiotics *in vivo* where metabolic transformation may give rise to chemical species responsible for the interaction with DNA, critical for neoplastic transformation.

Thus a considerable body of data now exists which supports the idea that DNA damage in somatic cells may be a critical event in the initiation of cancer, whilst damage that is transmitted through the germ cell line will have an effect on future generations and is therefore of concern to the population as a whole.

Evidence of genetic damage has been taken as sufficient to raise doubts as to the safety of a particular compound, and this concept of genotoxicity underlies many of the short-term tests developed over the last two decades. These tests are used not only to identify potentially hazardous compounds, but also for predictive value in risk assessment for synthetic chemicals by regulatory authorities.

By far the most common test-system is the 'Ames Test' or *Salmonella*/ microsome test (McCann *et al.* 1975, McCann & Ames 1976). This simple bacterial test uses a series of strains of *Salmonella typhimurium* which require histidine for growth and can be used in combination with a 9000 g liver supernatant fraction, a NADPH electron-generating system, and cofactors to screen for the mutagenic effect of an agent, by measuring reversion of these strains to histidine independence.

Many other test systems also exist (De Serres & Ashby 1981), and among those that have been developed are those which use the budding yeast *Saccharomyces cerevisiae* and to a lesser extent the fission yeast *Schizosaccharomyces pombe*.

S. cerevisiae is a unicellular organism which has a stable haploid and diploid stage, which is useful in so far as genetic effects may be monitored either during mitosis or meiosis and correlations made to somatic and germ line cells in higher eukaryotes.

The genetics and physiology of this yeast are well known, as a result a range of genetically well-defined, but simple, tests have been devised. These include: forward mutations (see Section 6.2.1.1) for base changes or deletions (von Borstel 1980); reverse mutations (see Section 6.2.1.2) for base substitutions or frameshifts (Zimmermann *et al.* 1975, Larimer *et al.* 1978), mutations in mitochondrial DNA (Schwaier *et al.* 1968); inter- and intragenic recombination (Zimmermann 1973, Zimmermann & Schwaier 1967), and chromosome aneuploidy during both mitosis (Parry & Zimmermann 1976) and meiosis (Sora *et al.* 1982). Differential killing in repair deficient strains and the induction of illegitimate mating between cells of the same mating type resulting from 'mating-type switch' events have also been used as assays for genotoxic compounds, but to a lesser extent than other genetic endpoints.

The use of *S. cerevisiae* combines the advantages of rapid cell division and simple culture methods (which have made the bacterial assays popular) with the examination of genetic systems more like those found in mammals. *S. cerevisiae* also possesses a cytochrome P-450 system (Lindenmayer & Smith 1964), the importance of which will be reviewed later.

Although *S. cerevisiae* is the most commonly used yeast in genotoxity assays, the fission yeast *S. pombe* is also used, albeit on a more limited basis. *S. pombe* has a stable haploid phase, and since Egel (1973) obtained a mating-type allele, *mei* 1–102, it also has a stable diploid phase, so that either the haploid or diploid cell response to mutagens can be investigated. The most common test systems used with *S. pombe* are point mutation assays (Loprieno 1973, Heslot 1962, Loprieno & Clarke 1965).

6.2 TEST SYSTEMS AND THEIR GENETIC ENDPOINTS

6.2.1 Point mutation

6.2.1.1 Forward mutation

The use of forward mutation assays are theoretically preferable to those assays employing reverse mutations as a genetic endpoint, in so far as many types of genetic alteration within a given gene can be detected. However, forward mutation tests do not usually permit the use of selective conditions to distinguish mutants from the remaining cell population and therefore greater numbers of colonies and more time to score mutational events are needed.

Two approaches have been used to detect forward mutation in haploid yeast strains. The most commonly used has been one based on the accumulation of a red pigment in strains with blocks in two reactions in adenine biosynthesis. Such strains carry mutant alleles of the genes *ADE* 1 or *ADE* 2.

Forward mutations are detected by mutation(s) at any of five genes which precede the production of the red pigment and consequently give rise to the appearance of white mutants among a background of red colonies (Roman 1956). This has been adopted by Marquardt *et al.* (1966) for mutagenicity testing. Sectored colonies allow the expression and subsequent detection of those mutations which were delayed through the processing of a lesion in the DNA to a new genotype or phenotype.

A second system for forward mutation assays is based on resistance to various different substances. For example, canavanine resistance is based on a defect in arginine permease (Lemontt 1977). Cells defective at this locus (*CAN* 1) are unable to take up canavinine from the medium and hence are resistant to its toxic effects. Another selection system is based on resistance to allyl alcohol, the genetic basis of which is a defect in the synthesis of alcohol dehydrogenase (Ciriacy 1975). Although the physiological and genetic basis of these selected traits are known, neither system has been well validated with chemical agents, a criticism which can also be made of the measurement of induced cycloheximide resistance in the strain *MP*1 (Fahrig 1975), where the biochemical processes are even less well understood. A disadvantage of many forward mutation assays is the nonselective methods that are used to identify mutational events. However, the strain *PV*1 has the genotype:

α *leu* 2-2, *pho* 1-100, LYS^+ ,

and forward mutations resulting in auxotrophy for lysine may be detected by plating on selective medium containing α-aminoadipate instead of ammonium sulphate (Parry *et al.* 1985).

In *S. pombe* two systems for detecting forward mutation have been used extensively. The first consists of a mutation from wild-type to adenine-requiring colonies and can be screeened for by the appearance of red colonies among a background of white colonies. This has been found to be due to mutations at the *ADE* 6 and *ADE* 7 loci (Leupold 1955, 1957).

The second system consists of a mutagen from adenine-requiring purple mutants (*ade* 6 or *ade* 7) to double-requiring white mutants; white colonies or sectors represent an additional mutation at the *ade* 1, *ade* 3, *ade* 4, *ade* 5, or *ade* 9 locus. These markers are carried in the mutator strain *P*1 (Loprieno 1973), which also carries the *rad* 10–198 mutation which shows increased sensitivity to chemical and physical mutagens.

6.2.1.2 Reverse mutation

Reverse mutation tests allow the use of selective conditions and are based on restoration of function, more usually restoration from auxotrophy (growth dependence on a specific nutrient) to prototrophy (growth independence). This restoration can be due to an exact reversal of the original defect in a gene coding for a required enzyme. Alternatively, a secondary mutation within the gene may lead to compensation of the original defect. In either event such changes usually require specific genetic alterations and are prone to mutagen specificity. To a certain extent this may be overcome by using a number of test-strains having different base-pair sequences in the gene studied, or a range of different genes and therefore representing different mutational spectra.

The latter approach has been adopted by Mehta & von Borstel (1981), who have constructed a strain *XV 185–14C* whose genotype is:

a ade 2-1, *arg* 4-17, *lys* 1-1. *trp* 5-48, *his* 1-7, *hom* 3-10 .

The markers for ochre nonsense mutations *ade* 2-1, *arg* 4-17, *lys* 1-1, and *trp* 5-48 are revertable by base substitution mutagens, or ochre supressor mutations in t-RNA loci, *hom* 3-10 is thought to be a frameshift defect based on its response to a number of known mutagens, while *his* 1-7 is a missense mutation which is reverted mainly by second site mutations.

The diploid yeast strain *D*7 described by Zimmermann (1975) and the strain *D*7 -144 (a derivative of *D*7 selected for its inability to undergo sporulation) have been developed as a multipurpose system for the detection of a variety of biological and genetic endpoints. The induction of nuclear point mutations at the homozygous *ilv* 1-92 allele is measured by the production of isoleucine-independent prototrophs on isoleucine-deficient medium. However, the range of mutagens to which this marker is susceptible is limited and is not regarded as a comprehensive screen for all environmental chemicals.

The strains *D*6 (Parry & Zimmermann 1976) and *D*61 - M (Parry & Eckardt 1985) carry the mutations *ade* 2-40 and *ilv* 1-92 respectively and can be used for the assay of the induction of adenine and isoleucine-independent prototrophs. However, both strains were developed in the first instance for the detection of induced chromosomal aneuploidy and are not as well validated for use for mutation induction.

Differential nutritional requirements have been used for study of reverse mutation induction in *S. pombe*. They are *his* 52, *his* 7, *met* 4, *ade* 6, and *ade* 7 (Heslot 1962, Loprieno & Clark 1965).

6.2.1.3 Cytoplasmic mutation

In yeast cells some chemical carcinogens primarily attack the mitochondria (see section 2.6) which results in lesions in organelles affecting the activity of certain nuclear genes which are involved in plasma membrane/cell surface biogenesis. Wilkie & Evans (1982) have observed that such surface changes are analagous to those seen in neoplastic transformation in mammalian cells. In yeast a number of chemicals that intercalate with DNA interfere with DNA replication and induce 'petite' mutants. Such colonies are incapable of aerobic respiration and are characterized by their small size, their inability to grow in nonfermentable carbon sources such as glycerol, and their inability to reduce the dye triphenyl tetrazolium.

6.2.2 Differential killing in repair deficient strains

The spontaneous mutability of some strains of *S. cerevisiae* has been explained by presuming that the spontaneous mutations are repaired by a mutagenic pathway. That is, when a particular repair pathway is blocked, the lesions that normally would have been repaired by that pathway may be channelled along another, normally competitive, but more mutagenic pathway (Brendel *et al.* 1970, Game & Cox 1972).

Knowledge of these mechanisms of DNA repair and their involvement in mutagenesis has come primarily from the genetic and biochemical studies of the effects of UV radiation on micro-organisms (Hanawalt *et al.* 1978, 1979, Haynes *et al.* 1966), and this is the subject of more detailed discussion by Cooper & Kelly (this volume).

For their use in yeast assays for genotoxicity testing certain strains have been constructed which carry radiation-sensitive mutations. Repair deficient, or radiation sensitive, mutants of *S. cerevisiae* were first reported by Nakai & Matsumoto (1967). Their discovery was closely followed by the isolation of many more such mutants which showed sensitivity to UV radiation (Cox & Parry 1968), X-irradiation (Resnick 1969, Strike 1978), methylmethane sulphonate (Prakash & Prakash 1977a), or other chemicals such as nitrous acid (Haynes & Kunz 1982). Some mutants also show cross-sensitivity to more than one of these mutagenic agents, for example *mms* 3 is sensitive both to MMS and UV (Prakash & Prakash 1977b), and *rad* 18 is sensitive to UV, X-rays (Resnick 1969) and MMS and nitrous acid (Brendel *et al.* 1970).

Haploid strains carrying individual radiation-sensitive markers, e.g. *rad* 50, *rad* 3 or *rad* 6 or strains carrying a combination e.g. *XS 1191–5D* : *trp* 2, *rad* 3, *rad* 18, *rad* 52, are used to assay the ability of the test compound to produce repairable DNA damage by measuring relative lethality as compared to wild-type strains.

6.2.3 Recombination

In diploid yeast cells mitotic and meiotic recombination can also alter the genome and this has been used to develop strains carrying markers where induced recombinational events can be measured. It is generally the mitotic events which are of practical value in genotoxicity studies and so these will be considered first.

6.2.3.1 Mitotic recombination

Mitotic recombination can be detected in yeast both inter- and intragenically. The first event generates cells homozygous for potentially detrimental genes, by reciprocal exchange, or crossover, between homologous chromosomes. It is generally assayed for by the production of recessive homozygous colonies and sectors produced in a heterozygous strain.

Mitotic gene conversion occurs between a pair of alleles differing at more than one site and represents the replacement of a sequence a few hundred nucleotides long on one chromosome, by the corresponding sequence of the homologous chromosome (Zimmermann 1971). It is characterized by its failure to yield reciprocal products during recombination (Yost *et al.* 1967). Gene conversion may be detected by the same selective techniques used for reverse mutation, that is by the production of prototrophic revertants.

Of particular significance in terms of genotoxicity assays is that both events described above are free of the mutagen specificities which hamper other assay systems. Both events are elevated by exposure to mutagens irrespective of their mode of action, and hence reflect the cell's response to a wide spectrum of genetic damage, but not necessarily at the damaged site itself (Fabre & Roman 1977).

A number of strains have been constructed for use in assaying environmental chemicals with respect to these genetic and endpoints. They are *D*3, *D*4, *D*5, *D*7 - 144, *JD*1 *D*6, and *D*61 - *M*, although the latter two are not commonly used.

The diploid strain *D*4 (Zimmermann 1975) has the genotype:

CHROMOSOME III	a / α	CHROMOSOME XII	gal 2 / GAL 2
CHROMOSOME XV	ade 2-2 / ade 2–1	CHROMOSOME VII	o leu 1 trp 5-12 / o LEU 1 trp 5–27

It may be used to assay for prototrophic colonies which have arisen through mitotic gene conversion, using selective medium lacking either tryptophan or adenine, e.g.

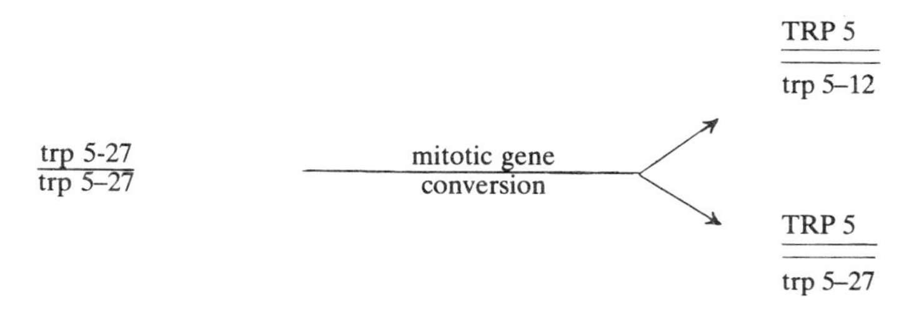

This strain has been used extensively to test a number of genotoxic chemicals (Zimmerman *et al.* 1984) but is subject to relatively high spontaneous reversion frequencies at the *ADE* 2 locus, and cultures have to be reselected before use.

A more useful strain is *D*7 (Zimmermann 1975). It carries the same heteroallelic *trp*5 marker as *D*4 and undergoes mitotic gene conversion in the same way as above, but the *ade*2 genotype is different and may be used to monitor mitotic crossing over. Its genotype is:

CHROMOSOME III	a / α	CHROMOSOME VII	o trp 5-12 cyh^{r}_{2} / o trp 5-27 CYH^{s}_{2}
CHROMOSOME XV	ade 2-40 / ade 2-119	CHROMOSOME V	ilv 1-92 / ilv 1-92

ade 2-40 is a completely inactive allele af *ADE* 2 which produces deep red colonies, whereas *ade* 2-119 is a leaky allele causing the accumulation of only a small amount of pigment, and this produces pink colonies. The original heteroallelic condition *ade* 2-40/*ade*2-119 forms white colonies, but mitotic crossing-over may give rise to cells homoallelic for the *ade* 2 mutations and give rise to red or pink colonies as illustrated below:

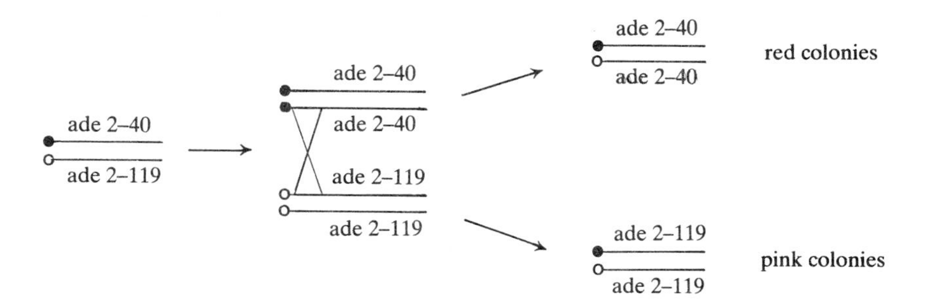

Mitotic crossing-over may also be assayed in *D*7 by the use of the recessive cycloheximide resistant cyh^r_2 allele on chromosome VII as illustrated:

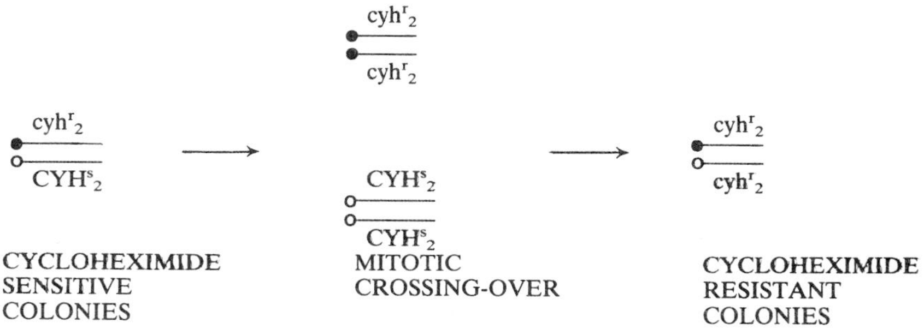

cyh^r_2
CYH^s_2

CYCLOHEXIMIDE
SENSITIVE
COLONIES

cyh^r_2
cyh^r_2
CYH^s_2
CYH^s_2

MITOTIC
CROSSING-OVER

cyh^r_2
cyh^r_2

CYCLOHEXIMIDE
RESISTANT
COLONIES

Crossing-over between CYH_2 and the centomere results in colonies which are capable of growth on medium containing cycloheximide (Kunz *et al.* 1980). Alternatively, the same marker may be used in the strains *D*6 and *D*61–M to measure mitotic crossing-over.

The strain *JD1* is 'petite' (Davies *et al.* 1975) and has the genotype:

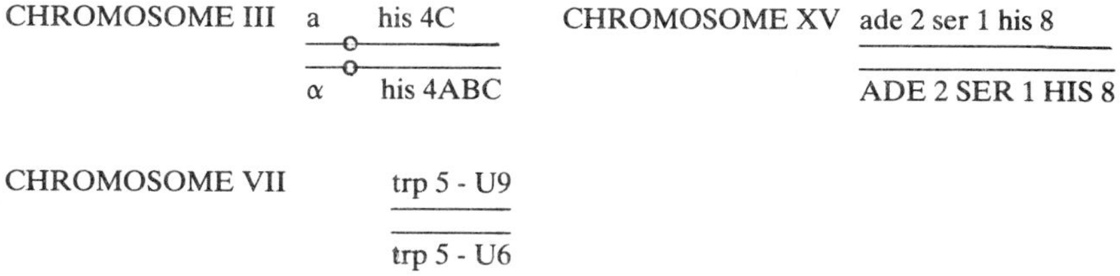

CHROMOSOME III　a　his 4C
　　　　　　　　α　his 4ABC

CHROMOSOME XV　ade 2 ser 1 his 8
　　　　　　　　ADE 2 SER 1 HIS 8

CHROMOSOME VII　trp 5 - U9
　　　　　　　　trp 5 - U6

It is heteroallelic at the *trp*5 and *his*4 leading to a requirement for tryptophan and histidine. Mitotic gene conversion may be scored by reversion to prototrophy at either or both loci.

Mitotic crossing-over may be assayed by the production of red colonies, or sectors homozygous for *ade*2 and the markers distal on chromosome XV (Parry *et al.*, 1984). as illustrated below:

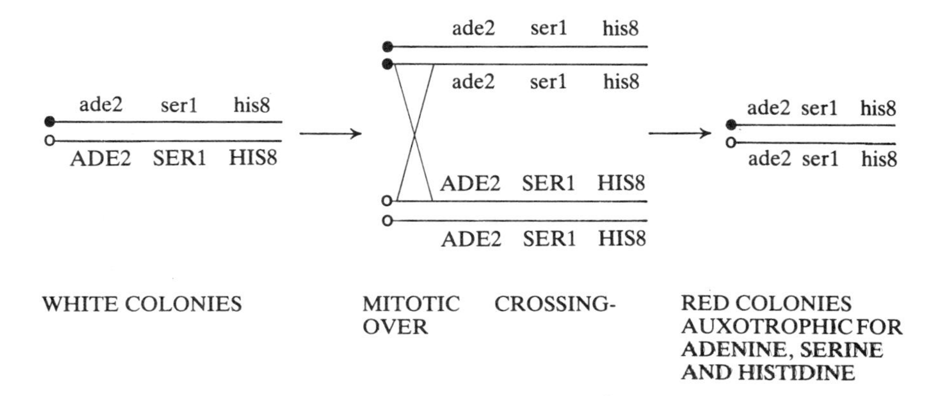

This strain and the strain *D7*-144 have been selected for their lack of ability to sporulate (however, it should be noted that under optimized conditions *D7* can be readily induced to sporulate, see later); both strains are suitable for long treatment times and an internal control for meiosis can be observed by an increase in coloured sectored colonies on control plates. If the meiotic pathway is entered this can lead to greatly elevated recombination levels.

6.2.3.2 Meiotic recombination

Induced meiotic effects by genotoxic chemicals and UV and X-ray irradiation may also be observed under certain conditions of growth, but the assays are by no means routinely employed and are not well validated.

Essentially two approaches may be adopted. The first is a measure of inhibition of meiosis and sporulation. For treatment at different stages of meiosis, cells are removed from sporulating cultures and resuspended in sporulation medium containing the chemical being tested. Sporulation frequency and plate counts for survival and induced recombination are determined. In this way the time of release from chemical sensitivity can be investigated.

The second approach involves a media-switch where calls are plated onto vegetative medium at different times in meiosis. In this way the timing of events leading to chemical sensitivity can be investigated. Exposure to the test chemical may be in the vegetative medium in the sporulation medium, or just prior to plating.

In theory any diploid yeast strain may be prompted to undergo meiosis given the correct growth conditions. However, even under the best conditions meiosis and sporulation do not take place in all the cells, and certain strains give higher levels than others, e.g. *D7* gives 70–80% sporulation by 36h (Kelly 1982). Using this strain the genetic endpoints of crossing-over and gene conversion during meiosis may be scored in the same way as for mitotic cells.

6.2.4 Aneuploidy

Evidence of genetic damage has been taken as sufficient to raise doubts as to the safety of a particular compound. However, while the concept of genotoxicity is clearly applicable to mutagens, it is not necessarily so for chemical carcinogens, and tests based solely on mutation or DNA damage as endpoints are not suitable for detecting all types of carcinogens.

For example, numerical chromosome aberrations can result in a variety of events. If they are due to simple chromosome damage the target is chromatin or centromere regions of the chromosome. The critically important mechanisms are a malfunction of the spindle fibre apparatus and possibly other factors involved in meiotic chromosome pairing, where the molecular targets are no longer DNA alone, but components of a different chemical nature.

Recommended tests in UK guidelines fall short of detecting non-disjunction events which are important in some genetic diseases, e.g. trisomy of chromosome 21 in Down's syndrome. Furthermore the observation by Parry *et al.* (1981) that certain tumour promoters induce mitotic aneuploidy, but are inactive in gene mutation assays (Trosko *et al.* 1977), highlights the relevance of such an assay system within a battery of tests.

6.2.4.1 Mitotic aneuploidy

Mitotic numerical aberrations can be detected in the strain *D*6 (Parry & Zimmermann 1976) and *D*61-M (Zimmermann, unpublished).

The genotypes of both strains are shown below:

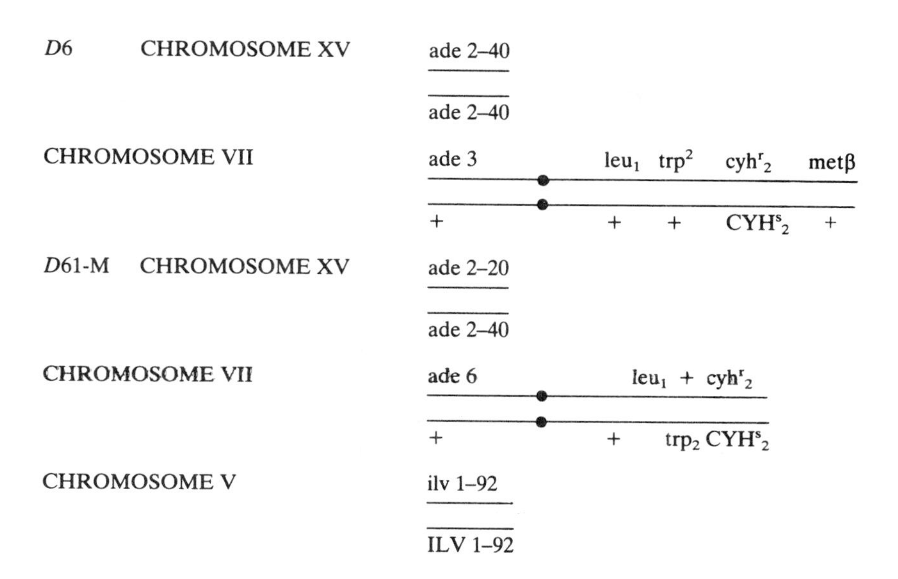

Both strains carry a series of recessive markers on chromosomes VII. They produce red colonies of solid complete medium low in adenine, owing to the presence of defective alleles *ade* 2-40 of the gene *adenine*-2 in a homozygous condition on chromosome XV. They do not form colonies on medium containing 2 mg/l cycloheximide. Mitotic chromosome non-disjunction leads to monosomy of chromosome VII, which gives rise to white cells owing to the presence of the defective allele of the gene *ade*-3 for *D*6 and *ade*-6 for *D*61-M; which results in a block in adenine biosynthesis prior to the *ade*-2 gene, thus preventing the formation of the red pigment characteristic of colonies carrying *ade*-2 mutations. Monosomic colonies are also cycloheximide resistant owing to expression of the cyh^r_2 marker on selective media, and monosomy may be verified by testing for the expression of the remaining mutant alleles carried on chromosome VII by plating on appropriate selective medium.

6.2.4.2 *Meiotic aneuploidy*

Meiotic aneuploid events are screened for by selecting for disomics (n+1), and two strains have been characterized. The first is *DIS*13 (Sora *et al.* 1982) and has the genotype:

CHROMOSOME V

$$\frac{can^r_1 \quad URA_3 \quad HOM_3 \quad his_1 \quad ARG_6 \quad ilv_1 \quad TRP_2 \quad met_5}{CAN^S_1 \quad ura_3 \quad hom_3 \quad HIS_1 \quad arg_6 \quad ILV_1 \quad trp_2 \quad MET_5}$$

CHROMOSOME VII

$$\frac{cyh^r_2 \quad leu_1}{CYH^s_2 \quad LEU_1}$$

CHROMOSOME XV

$$\frac{ade_2}{ADE_2}$$

CHROMOSOME III

$$\frac{a}{\alpha}$$

During meiosis spores are produced which carry multiple auxotrophic requirements, as shown on chromosome V, and are unable to grow on minimal medium. However, spores disomic for chromosome V (n+1) and diploid spores (2n) are capable of growth on minimal medium, and their frequency may be determined by the use of appropriate selective medium. Such diploid spores may represent an arrest of cells at a stage peculiar to yeast cells, but may bear some relevance to the production of polyploid zygotes (McDermott 1975) if similar arrest occurs in man.

In addition, the strain D_9J_2 may also be used for screening for disomics, its genotype is:

CHROMOSOME VII

ade5 + aro2 + chy2 + leu 1 + ade3

\+ lys5 + met13 + trp5 + ade6 +

The diploid is heterozygous for recessive canavanine resistance and so this may be used as a selective measure against diploid vegetative cells after plating on solid minimal medium and canavanine. Genetic analysis of prototrophic colonies arising on this medium indicates their disomic nature (Parry *et al.* 1979a, b).

6.2.5 Illegitimate mating

This is not a routine test, and has been included for completeness rather than its efficacy in testing for potential mutagens.

Mating-type 'switch' events from the α to the *a* mating-type can be induced in the presence of test compounds. Two strains can be used; their genotypes are:

*PV-4*a: α *his*7-1, *his*4-B26, *leu*2-1, *lys*2-A37, *met*A1

*PV-4*b: α *ade*1-6, *ade*2-163, *leu*2-2, *thr*4-B15, *lys*2-A12

Both strains carry a range of genetic markers conferring multiple auxotrophic requirements which prevent growth of the strains on selective medium.

Mating-type switch events result in the conversion of a proportion of the α mating type genes to the *a* mating type. If such an *a* cell is induced in strain *PV-4b* then mating may occur with α cells of *PV-4b* to produce a prototrophic diploid capable of growth on minimal selective medium. The frequency of such prototrophic diploids may be used to estimate the frequency of induced mating-type switch events.

6.3 CONSIDERATION OF CRITICAL FACTORS IN EXPERIMENTAL PROCEDURES

6.3.1 Handling the test compound

Investigators undertaking work with potentially mutagenic and/or carcinogenic compounds are advised to take into consideration their safe handling. Several published guidelines exist to this purpose (e.g. IARC 1979, MRC 1981). Similarly recommendations concerning preparation of test compounds have been given (Venitt *et al.* 1983).

6.3.2 Solvents

In yeast test systems trial compounds are more usually dissolved in the solvent dimethylsulphoxide (DMSO) – up to 5% v/v if they are not readily soluble or stable in aqueous solution. However, ethanol and acetone can also be used. At higher concentrations DMSO will induce mitotic gene conversion (Callen & Philpott 1977), and this should be borne in mind for any solvent and appropriate solvent controls set up.

6.3.3 Control chemicals

It is advisable for each assay to include, apart from the solvent control, reference chemicals for which historic data are available within the test system used. This allows the investigator to determine whether the assay has performed well and worked. In yeast assays the compounds cyclophosphamide and benzo(a)pyrene are two such examples, but a more comprehensive list is available (De Serres & Ashby 1981). When an assay includes a test for metabolic activation (see section 6.5) then a suitable control is also required. This usually takes the form of the compound being tested in the absence of the metabolic-mix to ensure there is no introduction of artefacts by the activation system. However, the intrinsic competence of the metabolic system is less easily defined, and as a result not readily monitored, hence the importance of including suitable reference chemicals as positive controls and the use of historic data.

6.3.4 Dose levels of the test compound

No recommendations exist for standard concentration increments which depend entirely on how rapidly the cells are inactivated, and no *a priori* valid upper concentration limit has been established for most chemicals. Ethyl methane-sulphonate, one of the most powerful yeast mutagens, is used at 10–20 mg/ml for treatment periods of 1–2 h, and several other potent direct-acting mutagens fall into this category, including sodium nitrite, hydroxyurea, dimethylhydrazine, and *N*-methyl-*N*-nitroso-guanidine (MNNG).

For other compounds there is no point in exceeding solubility limits, because absolute dose is not quantifiable under these conditions; and further increases in dose may only be achieved by extending treatment times. However, this latter approach has the concomitant problem of sporulation occurring and so this limits exposures to about 38 h, unless petite or haploid strains are used.

Some chemicals are toxic above a sharp threshold and positive mutagenic responses are observed within a narrow concentration range (Callen *et al.* 1980) and would be unobserved unless very small increments in dose were not used. It is therefore important to test chemicals at several concentrations below those including lethality.

6.3.5 Genetic stability of the test system

Each type of genetic event occurs at a typical rate, and this rate is reflected in a certain frequency of colony types which indicate the genetic change. It is important therefore, to determine the quality of the cell culture used and to check spontaneous reversion rates and induced reversion rates with reference chemicals against historical controls (as in 6.3.3.). Parry *et al.* (1984) have stressed the importance of this with respect to assays for mitotic gene conversion where continuous selection for low background may result in cultures that no longer consist of heteroallelic cells, but of homoallelic cells. These cells may not be revertable at all, or mutagen specific resulting in a much lower frequency of induction consistent with reverse mutation.

Bearing these points in mind it is of particular importance that permanent master-cultures of tester-strains should be stored in liquid nitrogen, or below −70°C. Working cultures should be prepared from a master-culture, never by passage from a previously used working culture.

Unusually high control levels with diploid strains may also indicate that sporulation has occurred (see 6.3.4). This may be checked by microscopic examination for the presence of asci, although meiotic recombination levels can be achieved without spore formation. The use of 'petites' as an alternative to get round this problem must also be viewed with some caution, since it has been shown that some petites are more sensitive to chemicals than the original 'grande' (Zimmermann 1969, Mayer *et al.* 1976, Siebert *et al.* 1979).

6.3.6 Precautions with selective procedures

Apart from the problem of sporulation with extended treatment times there is also the problem of possible selection of spontaneous prototrophs over induced prototrophs. This may be associated with lowering pH values during growth on fermentable media such as glucose. The uptake of amino acids, pyrimidines, and purines is pH dependent. If a culture has reached low pH, auxotrophic cells have difficulty absorbing the required nutrients, whereas spontaneous revertants can continue to grow. The possibility of selective growth of prototrophs has to be considered when positive results are observed in cultures that have reached stationary-phase after long treatment times, but not in cultures still in log-phase. This may be overcome by suitable buffering of cell suspensions during treatment (6.4.2 and 6.4.3).

Selection can also occur through differential killing of non-revertant cells and has to be considered when there is no increase in the number of revertant colonies per plate, but only an observed increase in the frequency of revertants per survivors. Reconstruction experiments which compare the toxicity of the chemical in normal cells and in cells derived from a revertant clone will identify whether selection has occurred. If the toxicity is the same between cell types then selection can be discounted.

6.4 EXPERIMENTAL PROCEDURES

6.4.1 Introduction

Although no single recommended assay exists for yeast tests, either in stationary or log-phase treatments, there are certain accepted protocols which find wider use than others, and it is the purpose here to present the essential features of these. A more detailed step-by-step approach has been given by Parry & Parry (1984), which also includes recipes for media used.

It is desirable to measure both the lethal effects of a test substance as well as its mutagenic effect on cells so that a quantitative as well as qualitative interpretation of the data may be made. The most common approach is to expose cells to the test chemical in liquid medium and after an appropriate time wash the cells free of the chemical and plate them on complete medium to assess cell survival and on minimal medium supplemented with the appropriate amino acids to assess the induction of the genetic endpoint under test. The data are usually expressed as mutants/revertants per survivors, usually per 10^6 cells. This so-called 'treat-and-plate' method allows for assessment of both toxicity and mutagenic effects. While methodologies based on the classical *Salmonella* microsome test have been developed, i.e. yeast and test substance are mixed with top-agar and immediately poured onto the surface of the bottom agar (Brusick 1980), these are less sensitive than liquid treatments (Shahin & von Borstel 1978, Jagannath *et al.* 1981) and do not allow for assessment of viability.

Spot-tests and fluctuation tests based on those used with bacterial tester-stains have also been adopted for use with yeast (Brusick & Zeiger 1972, Parry 1971), but neither approach has been used for routine screening of test compounds.

6.4.2 Treatment of stationary-phase cells

Cells may either be grown to the end of log-phase in liquid growth medium or taken from agar plates. In either case the percentage budding is determined microscopically and should not exceed 5% of the individual cells. Cells grown in liquid medium are harvested by centrifugation (about 6000 rev/min for 5 minutes) and resuspended in phosphate buffer pH 7.0 (several times to completely remove any traces of growth medium). Cells taken from plates can be scraped off with a sterile spatula and resuspended in buffer to a cell density of about 2×10^7 cells/ml.

A convenient method of treatment that has been adopted for use with yeast is to use disposable plastic 30 ml universal bottles containing a final reaction volume of cells, test compound, and if used, S9, (6.5.3) of 2 ml., (Parry & Parry, 1984). The test compound is usually added in 100 μl aliquots of increasing doses. The samples are incubated at 28°C on an orbital shaker (cells have a tendency to settle to the bottom of tubes if a shaking water-bath is used), for 16–18 h for preliminary experiments using unknown test-compounds, or for less time if toxicity or genotoxic potency are high. Modifications to this approach are usually with respect to the test compound, e.g. the use of glassware instead of plastic since some compounds or their solvents may dissolve or react with the plastic.

Before plating, the reactions are stopped by harvesting the cells (this is facilitated by use of 30 ml universal containers) and washed with ice-cold isotonic saline until all traces of media are gone. With alkylating agents the reactions are stopped by washing with cold 10% sodium thiosulphate.

After washing, cells are resuspended in phosphate buffer pH 7.0 to a cell titre of about 1–2 × 10^7 cells/ml and plated on yeast complete medium to give between 100–200 colonies per plate and 1 × 10^6 cells/plate on minimal medium plus supplements when scoring for mitotic gene conversion, and at a higher cell density, about 1–2 × 10^7 cells per plate, for scoring reverse mutation.

Complete plates are incubated at 28°–30°C in the dark, and survival can be scored after 2–3 days, but selective plates take longer to grow: usually about 5 days for gene conversion assays, e.g. using the strain *JD*1 or *D*7, 6–8 days for mitotic crossing-over in the strain *D*7, and up to 10 days for mitotic aneupoloidy using strain *D*6.

6.4.3 Treatment of logarithmic-phase cells

The treatment regime for growing cells is essentially the same as that described for stationary-phase cells, but differs with respect to the medium in which the cells are treated.

Selected strains are grown on complete plates, and a single colony is resuspended in either phosphate buffer, pH 7.0, or isotonic saline. This is used as the initial inoculum for either yeast complete medium or yeast minimal medium supplemented with any growth requirements specific to the strain. If the colony has been taken from a stock plate, rather than a working culture, the cells may require a period of growth instead of resuspension into buffer to prepare the initial inoculum. This may be done by resuspending a single colony in 2 ml yeast complete medium in a 25 ml conical flask and incubating overnight at 28°C on an orbital shaker. 1 ml of this culture may be inoculated into 250 ml yeast complete or minimal medium + supplements and allowed to grow to an O.D. = 0.6 at 600 nm, or 1 × 10^7 cells/ml. This usually takes about 6 hours, but is strain dependent, and it is recommended (Parry *et al.* 1984) that the cell number and percentage budding be determined microscopically using a haemocytometer; alternatively, if available, a Coulter counter represents a more quantitative estimate of bud size and number relative to mother cells. Bud emergence has been shown to coincide with the onset of DNA replication (Hartwell 1974), and this is important for treatment with some compounds such as MNNG, which have been shown to act preferentially at the replication fork (Cerda-Olemedo *et al.* 1968).

Treatment in complete medium is thought by some workers to inhibit the action of the test compound by binding of the test compound to complex components of the medium such as yeast extract or peptone. A particular problem associated with the use of growth medium is the concomitant fall in pH to acid levels due to the excretion of acids into the medium by growing cells. The inhibition, or stimulation, of mutagenic activity under these circumstances is less easily controlled compared to stationary-phase treatments in buffer. However, several alternative treatment regimes may be followed to allow for this. These include treatment of cells in minimal medium + supplements which is buffered using a final concentration of 100 mM phosphate (Zimmermann & Scheel 1981); treatment of cells in a fifth or two fifths strength complete medium (Sharp & Parry 1918, Zimmermann & Scheel 1981); or treatment of growing cells in phosphate buffer (Callen & Philpot 1977, Callen *et al.* 1980, Parry 1982, Callen 1981). However, this latter procedure is only suitable for short treatment times, since the cells are usually only capable of one subsequent cell division in the buffer, or may arrest prior to this.

Treatment times and conditions of washing and plating are the same as for stationary-phase cell, but as mentioned earlier (6.3.5) long treatment times may lead to sporulation of diploid cultures, or selection of prototrophic colonies, and so it is essential that during treatments of growing cells the medium should not become exhausted, thereby allowing cells to enter stationary-phase. This may be more critical when diluted medium is being used.

6.4.4 Scoring genotoxicity

6.4.4.1 Mitotic gene conversion

Usually a minimum of three plates of each cell type are used for each treatment, the number of cells plated onto selective plates being adjusted to about 1×10^6 cells/plate so that a reliable count is obtained on the control plates, usually not less than 20, so that the spontaneous frequency may be calculated.

The number of colonies on the selective and complete plates are recorded and the conversion frequency, i.e. number of convertants per 10^6 survivors, is calculated. Some strains, e.g. *D7*, form clusters of cells during growth which may be dispersed prior to plating by vigorous agitation or by sonication. However, if this is not done, conversion frequencies may be expressed as convertants per colony-forming unit.

6.4.4.2 Mitotic crossing-over

Mitotic crossing-over is a well defined genetic event, and its detection in the strain *D7* has already been described (6.2.3.1). In this strain mitotic crossing-over is indicated by the appearance of colonies with a red and a pink sector on a medium with limited adenine (4 mg/l). Storage at low temperature (4°C) enhances formation of the red pigment in the colonies and facilitates scoring. Plates are scored for total colony count, red colonies and pink colonies, twin colonies, and red/pink sectored colonies. An increase in the frequency of red/pink sectored colonies is taken as evidence of mitotic crossing-over, requires the screening of large numbers of colonies, and is therefore not always practical for large routine screening procedures.

6.4.4.3 Point mutation

6.4.4.3.1 Forward mutation

The visual assay of white colonies among a background of red colonies is indicative of a point mutation at the adenine loci prior to the *ade*-2 block for certain strains, as has already been described (6.2.1.1). In practice a large number of colonies have to be screened, usually a minimum of 20 000 colonies (about 100 per plate) for compounds which show no activity (Brusick 1980). This problem may be relieved in part by plating cells on complete medium at much higher densities (about 1000 per plate) and screening colonies using a low-power microscope. However, although definitive, the test is not readily amenable to a routine testing procedure for large numbers of unknown compounds, so relatively few chemicals have been tested using this system.

The assay for canavanine resistance and allyl alcohol resistance have been described, and the major protocol difference with selection for resistance is the provision of a period of post-treatment growth under non-selective conditions. Without this, cells will still retain residual wild-type copies of the enzyme arginine permease and alcohol dehydrogenase, even those that have been mutated, and hence will be sensitive to the toxic effects of these substrates in the medium. The post-treatment growth period allows time for expression of the mutations; however, it also detracts from the 'user-friendliness' of the assay and neither test has been routinely employed.

6.4.4.3.2 Reverse mutation

Scoring for reverse mutation is essentially the same as for mitotic gene conversion. Total colony counts for prototrophic colonies on appropriate selective medium signify the induction of mutational events. Usually $1–2 \times 10^7$ cells per plate are required on selective medium so titres may have to be adjusted prior to plating. This is particularly the case with frameshift mutations, in the strains available, where low spontaneous events may not give reliable colony counts, and particularly high cell titres (typically 2×10^8 cells/ml) are required. Practically, this requires larger reaction volumes which may be concentrated prior to plating and greater care when scoring plates, since background colonies due to spurious 'feeding' effects may obscure true revertants.

6.4.4.4 Petite induction (see Chapter 2.6)

The most rapid and therefore suitable method for screening large numbers of colonies at various dose levels for petites is the method described by Ogur *et al.* (1957). The assay is readily incorporated into any test where the tester strain has white colonies (e.g. *D*7) and survival plates have aready been scored. As such it is a useful addition to other endpoints already being tested in a given strain.

The assay is based on the inability of petite cells to metabolize 2,3,5-triphenyltetrazolium chloride (TTC). An overlay of soft agar buffered at pH 7.0 and containing 0.1% TTC (added after autoclaving, otherwise TTC is reduced) is poured on top of complete plates after 2–3 days growth and after plates have been scored for survival. After 2–3 hours at room temperature the plates may be scored. 'Grande' or normal colonies are red in colour because they can metabolise TTC, but petites remain white. Petite frequency is usually expressed as % petite colonies per survivors.

Alternatively, the inability of petites to grow on non-fermentable carbon sources, such as glycerol, may be used as an assay. Cells have to be plated on complete medium and after 2–3 days growth scored for survival. Replica-plating onto glycerol plates identifies those colonies which are petites on the master plate. However, the method is tedious when applied to large number of plates and not useful for routine screening programmes.

6.4.4.5 Treatment of cells at specific stages in the cell-cycle

6.4.4.5.1 The mitotic cell-cycle

The mitotic cell-cycle has been extensively studied in *S. cerevisiae* and in *S. pombe* because of the ease of recognition and separation of the various stages of growth and because of the ease of genetic analysis.

Its primary use in any screening programme is not to determine whether a test compound is genotoxic or not, but rather to assess the activity of a compound as particular stages of the cell-cycle, i.e. G_1, S or G_2. Use is made of synchronized cultures, or more conveniently by separation of exponential-phase cells on a zonal-rotor (Davies *et al.* 1978) or preferably by centrifugal elutriation (Elliot & McLaughlin 1978).

6.4.4.5.2 The meiotic cell-cycle (see Chapter 4, section 11.6)

The process of sporulation in *S. cerevisiae* involves both meiosis and the formation of four spores in an ascus. It occurs only in diploid cells which are heterozygous for mating-type, i.e. *a*/α and only under appropriate conditions, e.g. it can be induced by transferring cells from growth medium to nitrogen-limited starvation medium.

A common method of analysis of meiotic events has involved the use of media shift from sporulation medium to vegetative medium during the course of meiosis right up to ascus formation. This method allows the analysis of early and later meiotic events followed by a return to mitotic growth and has been adopted for use as a protocol for studying genotoxicity during the meiotic cell-cycle in yeast.

Sporulation can take a variable amount of time depending on the particular strain used and can be anywhere between 7.5 and 30 h. Furthermore, the extent of sporulation within a culture is also variable between strains, particularly those incorporating multiple genetic markers, which is true of many tester strains.

Growth in an acetate presporulation medium (Fast 1973) to a cell concentration of 2×10^7 cells/ml, followed by incubation in a potassium acetate sporulation medium (von Borstel 1978) at the same cell concentration, produces a rapid and synchronous meiosis (Kelly 1982). For the strain *D*7 higher cell concentrations were found to delay the onset of meiosis and result in less synchrony (Kelly 1982). Under these conditions asci appear in sporulating cultures of *D*7 at about 16 h and increase to between 70% and 80% sporulation by 48 h,whereas in the supersporulating strain *SK*1, asci first appear at about 7.5 h into sporulation and increase to almost 100% by 9.5–10 h.

Several alternative approaches may be used and their differences lie not in the treatment, but the timing of treatment.

Mutagen treatment throughout sporulation
In the first instance cells may be harvested from presporulation medium and inoculated into sporulation medium for 2 h before treatment with the test compound. This period represents a compromise between allowing cells to complete mitosis before treatment of meiotic cells and the possibility that some agents might act at early times of meiotic development.

2 ml of sporulating culture at 2×10^7 cells/ml sterile containers are treated with increasing doses of the test compound (where the solvent is DMSO, this should not exceed 2% v/v). Treated cultures are incubated for 48 h at 28°C on an orbital shaker (200 rev/min). The volume of the reaction vessel is important, and for all sporulating cultures the flask volume should be at least ten times greater than the culture volume to ensure good aeration. (Parry & Parry. 1984).

After treatment, cells are washed three times in isotonic saline and then plated on complete medium and selective medium to determine viability and the genetic endpoint under test. Microscopic examination of each sample is also carried out to determine the percentage sporulation and any inhibitory effect the test compound may have had.

In some instances, mutagen treatment of 2 ml samples is done in phosphate buffer, pH 7.0, instead of sporulation medium for periods as short as 30–40 minutes for alkylating agents, up to 48 h immediately after harvesting from presporulation medium and/or at one hourly intervals for the first 12 h in sporulation medium. In this way cells are treated during a residual mitosis for comparison to the effect of treatment during meiosis; the sensitivity of different specific-stages during meiosis to the test compound may also be determined.

Meiotic aneuploidy may only be scored following the completion of sporulation and analysis of spore products for disomic spores. For this endpoint, using either *DIS*13 (2.4.2) or D_9J_2 (2.4.2), Parry & Parry (1984) recommend treatment of cells in phosphate buffer pH 7.0, immediately after transfer from presporulation medium to sporulation medium and at one-hourly intervals for the first 12 h in sporulation medium. Treatment times are short, 30–45 minutes, after which cells are washed free of the test compound and resuspended in the sporulation medium from which they were originally growing; i.e. *not* fresh medium, since the pH of the medium changes with sporulation and release into fresh medium may alter the timing of sporulation for a particular sample. Samples are incubated at 28°C on an orbital shaker for a further 30 h, or longer if treatment has delayed sporulation. Cells are then resuspended at a cell titre of 1×10^7 cell/ml in buffer pH 5.8 and ascus wall digestion carried out over 12 h at 28°C using either sterile helicase or mushroom extract to release the spores.

Spores are resuspended in isotonic saline at 1×10^7 spores/ml and separated either by vigorous agitation or sonication, and then plated on complete plates to determine viability and selective plates for the endpoint being tested.

The significance of meiotic yeast test systems is not readily assessed, but they do offer an opportunity to compare sensitivities to environmental agents between mitotic and meiotic cells which bear some relevance to differences between somatic and germ-line cells. It is unlikely that they would be used for routine screening, however, because of the length of time required for each protocol and analysis of data.

6.5 THE ROLE OF METABOLISM IN GENOTOXICITY (see Chapter 5, Section 2.2.2)

6.5.1 Introduction

Although DNA is established as the ultimate, although not obligatory, primary target of chemical mutagens (Drake & Baltz 1976), other factors within a cell can influence whether a compound will finally damage DNA; factors such as DNA repair or enzyme-mediated biostransformations of chemicals to more or less mutagenic species.

The indirect-acting nature of some classes of chemical has already been mentioned (6.1.1). Their ultimate genotoxic activity is dependent upon the presence of appropriate activating enzymes. Although promutagens (as these compounds are sometimes called) comprise a wide-ranging structurally unrelated class of compounds, their common feature is their lipophilic nature at physiological pH. As such they become substrates for reactions catalysed by enzymes that are part of the normal detoxification mechanisms of the cells. There are two principal phases of reaction: (i) metabolic transformation, and (ii) conjugation (Parke 1968), and only the essential features of these will be described. Briefly, metabolic transformations are reactions in which the foreign compound undergoes one or more of a wide variety of oxidations, reductions, and hydrolysis. These usually result in the introduction of functional groups which increase the polarity of the molecule and act as centres for a second phase of reaction. These phase I reactions usually take place in the liver and involve principally the mono-oxygenases. The enzyme systems involved are situated on the endoplasmic reticulum and are thus present in the microsomal fractions obtained when tissue homogenates are subjected to differential centrifugation, specifically the 9000 g postmitochondrial supernatant (or 'S9' fraction from homogenized liver).

Conjugations, or phase II reactions, are syntheses by which the foreign compound, or any of its metabolites, is combined with molecules or groups such as glucoronic/sulphuric acids, amino acids, methyl, and other alkyl groups. These generally make the molecule more polar and less lipid-soluble and therefore more readily excreted.

Certain lipophilic compounds lack a functional group and are essentially unreactive, e.g. Benzo(a)pyrene. In such cases an initial oxygenation reaction is required to introduce a functional group onto the compound for subsequent metabolic transformation. This is usually catalysed by membrane-bound (microsomal) mono-oxygenases which represent a complex system. This system has been resolved into three essential components: NADPH-cytochrome P-450 reductase, phosphatidyl choline, and a haemoprotein cyt. P-450 that functions as a terminal oxidase.

The role of cytochrome P-450 (see section 5.2.2.2) in the hydroxylation of drugs was elucidated in a series of studies by Cooper *et al.* (1965), and substrate specificity was found to involve a variety of liposoluble organic compounds. This together with evidence that cytochrome P-450 exists in multiple forms, together with the inducibility of the various forms of specific substrates, provides mammalian tissue with an extremely versatile system capable of catalysing the oxygenation of a diverse range of endogenous and xenobiotic substrates.

6.5.2 *In vitro* test systems

While cultured mammalian cells and bacterial cells can be used to detect genotoxic activity *in vitro*, a major problem has been that such sample models cannot accurately mimic the complex kinetics of foreign compounds *in vivo*. Although some microorganisms, including some bacteria, have been shown to possess a cytochrome P-450 dependent monooxygenase system which resembles mammalian cytochrome P-450 (e.g. *Pseudomonas putida* and *Rhizobuim japonicum*, Dus *et al.* 1974, 1976), the test organisms routinely employed in short-term tests are generally ineffective in activating mutagens to their ultimate genotoxic forms. This has led to the development of 'metabolic mixes' which may be added to test systems in order to mimic the *in vivo* metabolism of test compounds. Such preparations are from the S9 fraction (see above) from mammalian tissue, usually the liver of Aroclor-treated rats (Ames *et al.* 1975, McCann *et al.* 1975).

6.5.3 Exogenous activation in yeast assays

For general screening in yeast the widely accepted method of applying a S9-mix to the yeast and test substance is in liquid, in the treat-and-plate method described earlier (6.4.1). Parry & Parry (1984) recommend a protocol where 50 μl–150 μl of S9-mix are added to 1 ml yeast cells (either stationary or logarithmic phase) and 100 μl of the test compound in a suitable solvent, the whole being made up to a final volume of 2 ml with phosphate buffer pH 7.0 and incubated at 37°C for 2 h without shaking and then at 28°C for 16 h on an orbital shaker. The most common S9-mix used is that from Aroclor-induced rat livers, although variations in the proportions of cofactors and protein levels differ between laboratories (Loprieno 1981, Sharp & Parry 1981).

6.5.4 Endogenous activation in yeast assays (see Chapter 5)

The presence of cyotchrome P-450 in yeast has been demonstrated by Lindenmayer & Smith (1964), and in 1969 Ishidate *et al.* studied the haemoproteins of anaerobically grown yeast using a cell free preparation. They reported the presence not only of cyt. a, and cyt. c peroxidase, but the presence of at least three other haemoproteins, cytochrome P-450, cytochrome b, and a cytochrome P-420-like haemoprotein. At about the same time Yoshida & Kamaoka (1969) and Yoshida *et al.* (1974) identified the cytochrome b, as cytochrome b_5 and concluded that the particulate bound haemoproteins of anaerobically grown cells made up an electron transport system which resembled those of microsomes from mammalian heptocytes. The cellular content of the cytochrome P-450 in cells of *S. cerevisiae* has been shown to vary over a wide range depending on the growth conditions, depression or induction, and the stage of the life cycle (Wiseman *et al.* 1975a). The concentration of cytochrome P-450 in anaerobically grown cells falls rapidly on aeration and also at late stages of the cell cycle if grown on low glucose medium (1% w/v). If higher glucose concentrations (up to 20% w/v) in the growth medium are used then the concentration of cytochrome P-450 remains higher for a long period of the yeast growth phase, and the high glucose concentrations afford a marked protective effect against the loss of cytochrome P-450 (Blatiak *et al.* 1980). However, the amount of cytochrome P-450 eventually falls to zero once the glucose is depleted to a critical concentration. Cytochrome P-450 in yeast grown under conditions of high glucose in the growth medium has indicated an ability to metabolize biphenyl by hydroxylation to the 4-hydroxylated derivative (Wiseman *et al.* 1975b), and also to show some demethylase activity when reacted with N-ethyl morphine and aminopyrine (Wiseman *et al.* 1975c). The latest full review in this subject is published in this book in Chapter 5).

Furthermore, Callen & Philpot (1977) have shown that growth of *S. cerevisiae* on comparatively low glucose medium (2% w/v) and under aerobic conditions also leads to elevated levels of cytochrome P-450, and under these conditions the stains *D*4 and *D*5 are capable of metabolic transformation of certain chemicals to genotoxic species. This latter test protocol did not optimize the level of cytochrome P-450 present over long treatment times. However, an assay system utilizing growth of yeast in low (0.5 w/v) and subsequently high (20% w/v) glucose containing medium has demonstrated the ability to detect indirect-acting mutagens requiring both cytochrome P-450 and cytochrome P-448 mediated activation (Kelly & Parry 1983). Cells grown under these conditions for the strains *NCYC*.240 (Wiseman & Woods 1977) and *D*6 and *JD*1 (Kelly & Parry 1983) have been shown to produce increased levels of cytochrome P-450, and the rate of disappearance of the enzyme is less than under aerobic conditions. This is of particular importance when treatment times take up to 18 h. Genotoxic responses of several reference compounds including cyclophosphamide, a cytochrome P-450 requiring compound (Hill *et al.* 1972) and benzo(a)-pyrene, a cytochrome P-448 requiring compound (Yang *et al.* 1978) using this latter approach, were to levels comparable to those obtained with an exogenous metabolic mix (Kelly & Parry 1983).

The protocol is based on a media switch of cells grown from an initial inoculum of 1×10^3 cells/ml to 1×10^7 cells/ml in 0.5% (w/v) glucose complete medium to 20% (w/v) glucose medium. Treatment of 2 ml aliquots of yeast cells in the medium follow the same treat-and-plate regime described in 6.4.3. Modifications of the original protocol delay the treatment of cells until 4 h and 6 h (Parry & Eckardt 1985a, b) after switching to 20% glucose medium so that cytochrome P-450 levels are already elevated at the time of adding the test compound (Parry & Parry 1984). Using a similar protocol Del Caratore *et al.* (1983) have demonstrated a genetic effect of styrene in yeast.

More recently the range of chemicals tested using this assay for several yeast strains has been extended (Kelly & Parry 1985, Parry 1985, Ali & Parry 1985, Mitchell & Gilbert 1985, Parry & Eckardt 1985a, b). The results illustrate the potential that this system offers with respect to replacing an S9-mix for certain test compounds.

6.5.5 Host-mediated assay

Some of the ulitimate mutagens and/or carcinogens are extremely short-lived, and the host-mediated assay is designed to bring the indicator cells into proximity with the site of activation within the mammal rather than attempt to mimic the *in vivo* metabolic pathway using an S9-mix. Yeast is suitable for this because it can be 'incubated' in the host animal for many hours. Cells are injected into the blood stream and recovered from various organs such as the liver, lung, or kidney (Fahrig 1975, Frezza *et al.* 1979). However, although this type of assay has given positive results, they were weak and the system has never been the subject of serious validation.

6.5.6 Factors affecting assays using metabolic activation

Two recurrent problems associated with the use of S9-mixes are: (i) the pretreatment of animals with microsomal enzyme inducers, e.g. Aroclor 1254, leads to an imbalance between activating and deactivating enzyme profiles. Such imbalance has been advanced as the explanation for the large increase in mutation frequency obtained with benzo(a)pyrene in *Salmonella typhimuruim* when a microsomal fraction from control rat livers was replaced by the corresponding fraction from Arochlor-treated rats (Rasmussen & Wang 1974). Thus manipulation of the activities of enzymes by pretreatment with inhibitors of deactivating enzymes, e.g. epoxide hydratase and/or pretreatment with microsomal enzyme inducers, can result in changes in mutagenicity within a given test system. A second factor to be taken into consideration when using S9's is the inherent difference between *in vivo* metabolism of a compound, which will generate reactive chemical metabolites within the cell and *in vitro* activation systems where the reactive chemical species is generated outside the cell. The latter situation will discriminate against compounds with short-half lives whose production outside the cell may reduce the probability of the metabolites reacting with the nuclear DNA. Hence the interesting pharmacokinetic approach of host-mediated assay, which could also lead to the identification of the sites of activation and to the distribution patterns of ultimate genotoxic species in the whole animal.

Although the problems discussed earlier (6.3.6) associated with the use of growing cells and treatment in medium rather than buffer have led to some workers' preference for work with stationary-phase cultures, the potential for endogenous metabolism in growing cells represents an opportunity for microbial cells to be a more adequate model of *in vivo* metabolism in whole animal. However, before this can be realized a more thorough understanding of the yeast's metabolic proficiency with respect to different classes of compound and the spectrum of metabolites yielded needs to be investigated, particularly for known chemical mutagens/carcinogens and compared to that of mammalian systems.

6.6 ANALYSIS AND INTERPRETATION

A problem with selective tests, such as those described here, is that what is observed experimentally is not the induction of a particular genetic endpoint such as mutation, but certain types of colonies that are thought to reflect the occurrence of such events. As such these assays are prone to artifacts, and careful consideration of experimental design and interpretation of the data must be made.

A general consensus is that a dose-related absolute increase in the type of colony reflecting a particular genetic event constitutes a positive result in genotoxicity assays. Equally important is survival, which indicates whether a particular treatment had any effect. Consequently, comparison between controls and samples has to be based on relative frequencies, which are numbers of genetically changed cells over survivors.

If the increase is only relative (i.e. the absolute numbers of genetic events decrease in response to dose) then there arise a number of questions concerning the mathematical and biological significance of such a result, and it is because of this that the appropriate analysis should be performed.

6.6.1 Quantitative analysis

Statistical analysis is limited by experimental design which may not allow sufficient data to be confidently analysed; for example, too few dose points, small numbers of replicates either in or between experiments, and/or inappropriate controls.

Parry *et al.* (1984) have recommended a minimum of five different dose levels which should be evenly spaced on a logarithmic scale and which take into consideration toxicity and/or maximum solubility. The number of replicates in an experiment is determined by the statistical analysis to be performed. Generally, triplicate plates at least are required for the '*t*-test' (Parker 1976) which evaluates differences between means based on normal distributions, while four or more replicates will be needed for nonparametric analysis.

In genetic toxicology the question is whether the treated sample exceeds (rather than differs from) the control, and so probability values calculated from one-sided tests are considered more appropriate than values for single comparisons by a two-sided test (Parry *et al.* 1984). The 'p'-value for a one-sided test is exactly half that for a two-sided test and is usually set either $p = 0.05$ or $p = 0.01$, although other values may be used. An indepth discussion of appropriate analyses is given by Parry *et al.* (1984).

A more sophisticated evaluation of genetic activity can be made by plotting the data as yield of gentic events, according to Haynes & Eckardt (1980). For a given gentic endpoint and exposure dose of the compound, the net induction of viable mutants (mutant yield) is calculated. When these values are positive the test agent is classified as showing inducing capacity, whereas negative values indicate that the test agent is genetically inactive under the experimental conditions. Furthermore, the integral under the yield curves, plotted against lethal hits (negative logarithm of the surviving fraction) for each exposure can be calculated. Eckardt & Haynes (1980) have shown that the yield integral (which reflects the total number of viable mutants produced over the entire dose range of an agent) can be used for quantitative comparison of the mutagenic efficiency of genotoxic agents. Positive numbers for the integral yield values indicate that at least some doses of the agent induce the genetic endpoint under consideration, and a yield of zero indicates no inducing activity.

However, a negative response does not necessarily reflect a lack of genetic activity, but may be due to insufficient testing procedures, such as the use of stationary versus growing cells. It is important therefore to consider whether a test has been performed adequately. The use and importance of control chemicals has already been described (6.3.3). An additional control to show that the exogenous activation system has worked is also necessary, and a control which is structurally related to the test compound is important in this respect for assurance that certain conditions of metabolic activation can be, or have been met.

6.6.2 Biological significance

It is the generally accepted view that test compounds which elicit a minimum of 1.5-fold increase in genetic activity over the control value are considered biologically significant. This value, however, should be in excess of the minimum detectable by statistical analysis, since quantitative significance (biological or genetical) in the absence of mathematical significance is unrealistic (Parry *et al.*, 1984).

6.7 VALIDATION AND HAZARD EVALUATION — AN OVERVIEW OF YEAST ASSAYS

Since November 1985 the EEC legislation on pharmaceuticals has required that all new active ingredients should be tested for mutagenic potential before maerketing. However, the predictive value of short-term tests with respect to mutagenic and/or carcinogenic hazard to man represents a considerable problem.

Since no one test system is recognized as being sufficient to detect all genetic endpoints (limitations to specific test-systems are described in the UKEMS guidelines for mutagenicity testing, 1984), the EEC committee for Proprietary Medicinal Products (CPMP) has developed a guideline on mutagenicity testing which recommends a battery of four tests selected from four different test categories (Directive 83/5701, 1984), and which has already been included in the latest UK guidelines for product application. A fifth category, not included specifically in the guidelines, is primary DNA damage/repair, and included in this section are yeast assays to measure induced mitotic recombination.

Given the spectrum of genetic endpoints which may be assayed for in yeast, it represents a useful organism with which to test compounds.

However, validation of protocols and strains is a problem, and this has perhaps been the main reason for the assignation of yeast assays to a supplementary test role.

In the absence of any rules for establishing that a given assay is validated a consensus has arisen which is largely based on the performance of the *Salmonella*/microsome test. Large numbers and different types of compound have been tested using this assay, and their relevance and reliability in detecting mutagens have been the subject of several large-scale trials (Ames *et al.* 1975, Purchase *et al.* 1978, McMahon *et al.* 1979, Rinkus & Legator 1979, IARC 1980, Bartsch *et al.* 1980, De Serres & Ashby 1981, Venitt 1982). It is true to say that this test system in particular has been studied, validated, characterized, and understood in far greater detail and by many more testing laboratories than most other short-term tests, including yeast assays.

Adopting the approach of testing large numbers of compounds as a criterion for validation, yeast assays have been recently subjected to evaluation in a number of trials (De Serres & Ashby 1981, Parry *et al.* 1984, Parry & Arlett, 1985, Parry 1985). The data from these studies support the view that for chemicals with the ability to interact with nuclear DNA the gene conversion endpoint is capable of detecting genetic activity in a non-specific manner without the requirement for a proliferation of different tester strains or selection at a variety of loci.

Using a different criterion for validation, i.e. to use given or similar strains in at least three different laboratories on the same weak, direct, and indirect mutagens and nonmutagens, the strains *D*4, *D*5, *JD*1, and *D*7 are all strains which are close to being validated.

In addition to their role in general screening, certain yeast assays may be chosen to fill specific gaps within a test programme, but without the need to set up new testing facilities or gain new expertise. For example, the yeast aneuploidy assays can be used to detect not only mutagenic compounds, but also other classes of agents which represent a potential carcinogenic hazard, such as tumour promoters and hormones such as diethylstilboestrol (Parry *et al.* 1981). The ability of the strain *D*6 to detect such additional nonmutagenic species represents an important supplementary test system that can be handled by conventional microbial techniques.

Apart from the need to detect those chemicals which present a non-mutagenic hazard, such as tumour promoters, another important area of study in genotoxicity is the metabolism of foreign compounds *in vivo* and the requirement to mimic this *in vitro*. To date, the most accepted solution to this problem is the addition of an S9-mix to the assay which may vary in preparation and storage from laboratory to laboratory (Venitt *et al.* 1983).

Refinement of existing yeast protocols offer the possibility to regulate the complex interaction of activating (e.g. cytochrome P-450) versus deactivating (e.g. the tripeptide glutathione) enzymes which catalyse the production of reactive electrophilic intermediates as well as reactive oxygen species which cause DNA damage.

A variety of approaches can be envisaged which would allow elevated levels of activating enzymes in tester strains. One of these is by induction of appropriate enzymes (see Chapter 3). Also, for example the expression of yeast or mammlian cytochrome P-450 on high copy number plasmids in tester strains and/or expression of cloned cytochrome P-450 under regulation of suitable high expression yeast promoters is a valuable advance (Oeda *et al.* 1985). Alternatively, with more detailed knowledge of the DNA sequence and understanding of the importance of domains within the protein, modified cytochrome P-450 with higher specific activity than the yeast ancestral protein (produced by protein engineering or protein modification) could be introduced into tester strains. Such approaches would still be the subject of extensive validation, but would offer a system of *in vivo* metabolism rather than *in vitro*, to more closely resemble the pharmaco kinetics of xenobiotics in higher animals.

The advance of recombinant DNA studies in yeast has also opened up the possibility of construction of new assays, which may be more rapid to perform and therefore more readily accepted on the basis of routine screening for large numbers of compounds. For example, the gene fusion technique which has identified six damage inducible (*DIN*) genes (Ruby & Szostak 1985) and which was used to study *RAD* gene regulation (see Cooper & Kelly, this volume) represents the potential for a simple test which would measure induced DNA damage by assaying for B-galactosidase activity, with a broad range of response to both physical and chemical mutagens in contrast to the promoters of the genes involved in the SOS response in *E. coli* which are induced after particular forms of DNA damage (Walker 1984).

Thus, the potential exists for developing existing yeast assays for which validation is already underway and for introducing new systems which may be selected to solve specific problems in test programmes using conventional microbiological techniques.

REFERENCES

Ali, F. & Parry, J. M. (1985). In: *Comparative genetic toxicology. The second UKEMS collaborative study*. J. M. Parry and C. F. Arlett (eds) Macmillan. pp. 253–258.

Ames, B. N., McCann, J. & Yamasaki, E. (1975). *Mut. Res.* **31**, 347–364.

Auerbach, C. (1947). *Exp. Cell, Res. Suppl.* **1** 93–96.

Bartsch, H., Malaveille, C., Camus, A.-M., Martel-Planche, G., Brun, G., Hautefeuille, A., Sabadie, N., Barbin, A., Kuroki, T., Drevon, C., Piccoli, C. & Montesano, R. (1980). In: *Molecular and cellular aspects of carcinogenic screening tests*. Bartsch, H., Montesano, R. and Tomatis, L. (eds) In: *IARC Scientific Publications*. No. 27, pp. 179–241.

Blatiak, A. A., Gondal, J. A., & Wiseman, A. (1980). *Biochem. Soc. Trans.* **8** 711–712.

Brendel, M., Khan, N. A. & Haynes, R. H. (1970). *Mol. Gen. Genet.* **106** 289–295.

Brookes, P. & Lawley, P. D. (1964). *J. Cell. Comp. Physiol.* **64** (Suppl. 1) 111–120.
Brusick, D. J. (1980). In: *Principles of genetic toxicology.* Plenum Press, New York, pp. 199–201.
Brusick, D. J. & Zeiger, E. (1972). *Mut. Res.* **14** 271–275.
Callen, D. F. (1981). *Environ. Mutagenesis.* **3** 651–658.
Callen, D. F. & Philpot, R. M. (1977). *Mut. Res.* **45** 309–324.
Callen, D. F., Wolf, C. R., & Philpot, R. M. (1980). *Mut. Res.* **77** 55–63.
Cerda-Olemedo, E., Hanawalt, P. C., & Guerda, W. (1968). *J. Mol. Biol.* **33** 705–719.
Ciriacy, M. (1975). *Mol. Gen. Genet.* **138** 157–164.
Cooper, D. Y., Levin, S., Narasimhulu, S., Rosenthal, O. & Estabrook, R. W. (1965). *Science (Wash. D.C.)* **147** 400–402.
Cox, B. S. & Parry, J. M. (1968). *Mut. Res.* **6** 37–55.
Davies, P. J., Evans, W. E., & Parry, J. M. (1978). *Mut. Res.* **29** 301–314.
Davies, P. J., Tippens, R. S., & Parry, J. M. (1978) *Mut. Res.* **51** 327–346.
De Serres, F. J. & Ashby, J. (eds) (1981). In: *Progress in mutation research*, Vol. 1. Elsevier/North Holland.
Del Carratore, R., Bronzetti, G., Baver, C., Corsic, C., Nieri, R., Paolini, M. & Gioagoni, P. (1983). *Mut. Res.* **121** 117–123.
Drake, J. W. & Baltz, R. H. (1976). *Annv. Rev. Biochem.* **45** 11–37.
Dus, K., Litchfield, W. J., & Miguel, A. G. (1974). *Biochem. Biophys. Res. Commun.* **60** 15–21.
Dus, K., Goewert, R., Weaver, C. C., Carrey, D., & Appleby, C. A. (1976). *Biochem. Biophys. Res. Commun.* **69** 437–445.
Echardt, F. & Haynes, R. H. (1980). *Mut. Res.* **74** 439–458.
Egel, R. (1973). *Mol. Gen. Genet.* **121** 277–284.
Elliott, S. G. & McLaughlin, C. S. (1978). *Proc. Natl. Acad. Sci. USA.* **75** 4384–4388.
Fabre, F. & Roman, H. (1977). *Proc. Natl. Acad. Sci. USA.* **74** 1667–1671.
Fahrig, R. (1975). *Mut. Res.* **31** 381–394.
Fast, D. (1973). *J. Bacteriol.* **116** 925–930.
Freese, E. (1963). In: *Molcular genetics*, Part 1. J. H. Taylor (ed). Academic Press, New York, pp. 207–269.
Frezza, D., Zeiger, E., & Gupta, B. N. (1979). *Mut. Res.*, **64** 295–305.
Game, J. C. & Cox, B. S. (1972), *Mut. Res.* **16** 353–362.
Hanawalt, P. C., Friedberg, E. C., & Fox, C. F. (eds) (1978). Academic Press, New York.
Hanawalt, P. C., Cooper, P. K., Ganesan, A. K., & Smith, C. A. (1979). *Ann. Rev. Biochem.* **48**, 783.
Hartwell, L. H. (1974). *Bacterial Reviews* **38** 164–198.
Haynes, R. H., Wolff, S., & Till, J. (1966). *Radiat. Res.* **6** 1.
Haynes, R. H. & Kunz, B. A. (1982). In: *The molecular bilogy of Saccharomyces cerevisiae.* Cold Spring Harbour.
Heslot, H. (1962). *Abh. Dtsch. Akad. Wiss. Berlin, Kl. Med.*, No. 1, 193–228.
Hill, D. L., Laster, W. R., & Struck, R. F. (1972). *Cancer Res.* **32** 658–665.

IARC (1979). Montesano, R., Bartsch, H., Boyland, E., Della Porta, G., Fishbein, L., Greisemer, R. A., Swan, A. B. and Tomatis, L. (eds) *IARC Scientific Publications*. No. 33.

Ishidate, K., Kawaguchi, K., Tagawa, K., & Hagihara, B. (1966). *J. Biochem.* **65** 375–383.

Jagannath, D. R., Vultaggio, D. M., & Brusick, D. J. (1981). In: *Progress in mutation research* Vol. 1. Evaluation of short-term tests for carcinogens. (F. J. de Serres and J. Ashby (eds). Elsevier/North Holland, New York, Amsterdam, Oxford.

Kelly, D. E. & Parry, J. M. (1983). *Mut. Res.* **108** 147–158.

Kelly, D. E. & Parry, J. M. (1985). In: *Comparative genetic toxicology*, pp. 211–220 Parry, J. M. and Arlett, C. F. (eds).

Kelly, S. L. (1982). PhD thesis (Wales).

Kunz, B. A., Hannan, M. A., & Haynes, R. H. (1980). *Cancer Res.* **40** 2323–2329.

Larimer, F. W., Ramey, D. W., Lijinsky, W., & Epler, J. L. (1978). *Mut. Res.* **47** 155–161.

Lemontt, J. F. (1977). *Mut. Res.* **43** 165–178.

Leupold, U. (1957). *Schweiz, Z. Allg. Pathol. Bakteriol.* **20** 535–544.

Leupold, U. (1955). *Arch. Julius Klaus-Stift. Vererbungsforsch* **30** 506–516.

Lindenmayer, A. & Smith, L. (1964). *Biochim. Biophys. Acta* **93** 445–461.

Loprieno, N. (1973). *Genetic* (Suppl). **73** 161–164.

Loprieno, N. (1981). In: *Progress in mutation research.* F. J. de Serres and J. Ashby. (eds). Elsevier/North Holland.

Loprieno, N. & Clarke (1965). *Mut. Res.* **2** 312–319.

Malling, H. V. (1971) *MUt. Res.* **13** 425–429.

Marquardt, H., von Laer, U., & Zimmermann, F. K. (1966). *Z. Vererbungsl.* **98** 1–9.

Mayer, V. W., Hybner, C. J., & Brusick, D. J. (1976). *Mut. Res.* **37** 201–212.

McCann, J., Choi, E., Yamasaki, E., & Ames, B. N. (1975). *Proc. Natl. Acad. Sci.* **72** 5135.

McCann, J. & Ames, B. N. (1976). *Proc. Natl. Acad. Sci.* **73** 950.

McDermott, A. (1975). Chapman & Hall Ltd., London, pp. 38.

McMahon, R. E., Cline, J. C., & Thompson, C. Z. (1979). Cancer Res. **39** 682–693.

Metha, R. D. & von Borstel, R. C. (1981). In: *Progress in mutation research.* Vol. 1. Evaluation of short-term tests for carcinogens. F. J. de Serres and J. Ashby (eds). Elsevier/North Holland, New York, Amsterdam, Oxford.

Mitchell, I. & Gilbert, P. J. (1985). In: *Comparative genetic toxicology. The Second UKEMS Collaborative Study.* J. M. Parry and C. F. Arlett (eds). MacMillan, pp. 241–252.

MRC (1981). MRC, London.

Muller, H. J. (1927). *Science* **46** 84–87.

Nakai, S. & Matsumoto, S. (1967). *Mut. Res.* **4** 129–136.

Ogur, M., St. John, R. & Nagai, S. (1957). *Science* **125** 928–929.

Parker, R. E. (1976). In: *Introductory statistics for biology; studies in biology* no. 43. pp. 20.
Parke, D. V. (1968). In: *The biochemistry of foreign compounds*. D. V. Parke (ed.). International series of monographs in pure and applied biology. Pergamon Press, Oxford.
Parry, J. M. (1971). *LAB 21–45*, pp. 417–419.
Parry, J. M. (1982). *Mut. Res.* **100** 145–151.
Parry, J. M. (1985). In: *Comparative genetic toxicology. The Second UKEMS Collaborative Study*. J. M. Parry and C. F. Arlett, (eds), MacMillan, pp. 231–239.
Parry, J. M., Sharp, D. & Parry, E. M. (1979a). *Environ. Health Perspect.* **31** 97–111.
Parry, J. M., Sharp, D., Tippins, R. S., & Parry, E. M. (1979b). *Mut. Res.* **61** 37–55.
Parry, J. M., Parry, E. M., & Barrett, C. (1981). *Nature* **294** 263–265.
Parry, J. M., Brooks, T., Mitchell, I. & Wilcox, P. (1984). In: *UKEMS subcommittee on guidelines for mutagenicity testing*. (Part IIA), pp..27–61.
Parry, J. M., Arni, P., Brooks, T., Carere, A., Ferguson, L., Heinisch, J., Inge-Vechtomov, S., Loprieno, N., Nestmann, E., & von Borstel, R. (1985). In: *Progress in mutation research* Vol. 5. Elsevier Science Publishers, Amsterdam, pp. 25–46.
Parry, J. M. & Eckardt, F. (1985a). In: *Progress in mutation research* Vol. 5. Ashby, J. & de Serres, F. J. (eds) Elsevier Sciences Publishers/Amsterdam.
Parry, J. M. & Eckardt, F. (1985b). In: *Progress in mutation research*. Ashby, J. and de Serres, F. J. (eds) *Vol. 5*. Elsevier Science Publishers/Amsterdam.
Parry, E. M. & Parry, J. M. (1984). In: *Mutagenicity testing: a practical approach*. S. Venitt and J. M. Parry (eds). IRL Press, Oxford.
Parry, J. M. & Zimmermann, F. K. (1976). *Mut. Res.* **36** 49–66.
Prakash, L. & Prakash, S. (1977a). *Genetics* **86** 33–55.
Prakash, L., & Prakash, S. (1977b). *Genetics* **87** 229–236.
Purchase, I. F. H., Longstaff, E., Ashby, J., Styles, J. A., Anderson, D., Lefevre, P. A., & Westwood, F. R. (1978). *Br. J. Cancer* **37** 873–959.
Rasmussen, R. E. & Wang, I. Y. (1974). *Cancer Res.* **34** 2290–2295.
Resnick, M. (1969). *Genetics* **62** 519–531.
Rinkus, S. J. & Legator, M. S. (1979). *Cancer Res.* **39** 3289–3318.
Roman, H. (1956). *Cold Spring Harbor Symp. Quant. Biol.* **21** 175–183.
Ruby, S. W. & Szostak, J. W. (1985). *Mol. Cell. Biol.* **5** 75.
Sasaki, T. & Yoshida, T. (1935). *Virchow Arch. (Path. Anat.)* **295** 175.
Schwaier, R., Nashed, N. & Zimmermann, F. K. (1968). *Mol. Gen. Genet.* **102** 290–300.
Shahin, M. M. & von Borstel, R. C. (1978). *Mut. Res.* **53** 1–10.
Sharp, D. C. & Parry, J. M. (1981). In: *Progress in mutation research*. Vol. 1. de Serres, F. J. and Ashby, J. (eds) Elsevier, North Holland. pp. 491.
Siebert, D., Bayer, U. & Marquardt, H. (1979). *Mut. Res.* **67** 145–156.

Sora, S., Lucchini, G., & Magni, G. E. (1982). *Genetics.* **101** 17.
Strike, T. L. (1978). PhD thesis, University of California, Davies.
Trosko, J. E., Chang, C., Yotti, L. P., & Chu, E. H. Y. (1977). *Cancer Res.* **37** 188–193.
Venitt, S. (1982). *Mut. Res.* **100** 91–109.
Venitt, S., Forster, R., & Longstaff, E. (1983). In: *UKEMS sub-committee on guidelines for mutagenicity testing.* Report, Part 1. Basic Test Battery. B. J. Dean (ed). UKEMS, SWANSEA. pp. 5–40.
Von Borstel, R. C. (1980). In: *Long-term and short-term screening assays for carcinogens. A critical appraisal.* IARC Monographs, Supplement 2. pp. 135–155.
Von Borstel, R. C. (1978). In: *Methods in cell biology.* Vol. **XX** (ed) D. M. Prescott, Acad. Press. pp. 1–24.
Walker, G. C. (1984). *Microbiol. Rev.* **48**(1): 60–93.
Watson, J. D. & Crick, F. H. E. (1953). *Nature* **171** 964–967.
Wilkie, D. & Evans, I. H. (1982). *Trends in Biochem. Sci.* **7** 147–151.
Wiseman, A., Lim, T.-K., & McCloud, C. (1975a). *Biochem. Soc. Trans.* **3** 276–278.
Wiseman, A., Gondal, J. A., & Sims, P. (1975b). *Biochem. Soc. Trans.* **3** 278–281.
Wiseman, A., Jay, F., & Gondal, J. A. (1975c). *J. Sci. Food. Ag.* **26** 539–540.
Wiseman, A. & Woods, L. F. J. (1977). *Biochem. Soc. Trans.* **5** 1520–1522.
Yang, S. K., Roller, P. P., & Gelboin, H. V. (1978). In: *Carcinogenesis.* Vol. 13. P. Jones and R. I. Freudenthal (eds), *Polynuclear Aromatic Hydrocarbons.* Raven Press, New York.
Yoshida, Y. & Kumaoka, H. (1969). *Biochim. Biophys. Acta* **189** 461–463.
Yoshida, Y., Kumaoka, H. & Sato, R. (1974). *J. Biochem.* **75** 1201–1210.
Yost, H. T., Chalett, R. S., & Finerty, J. P. (1967). *Nature* **215** 660–661.
Zimmermann, F. K. (1969). *Z. Krebsforsch.* **72** 65–71.
Zimmermann, F. K. (1971). *Mut. Res.* **11** 327–337.
Zimmermann, F. K. (1973). *Mut. Res.* **21** 263–269.
Zimmermann, F. K. (1975). *Mut. Res.* **31** 71–86.
Zimmermann, F. K. & Scheel, I. (1981). In: *Evaluation of short-term tests for carcinogens.* de Serres, F. J. and Ashby, J. (eds). Elsevier/North Holland. pp. 481–490.
Zimmermann, F. K. & Schwaier, R. (1967). *Mol. Gen. Genet.* **100** 63–76.
Zimmermann, F. K., Kern, R., & Rasenberger, H. (1975). *Mut. Res.* **28** 381–388.
Zimmermann, F. K., von Borstel, R. C., von Halle, E. S., Parry, J. M., Siebert, D., Zetterberg, G., Barale, R., & Loprieno, N. (1984). *Mut. Res.* **133** 199–244.

Index

acid phosphatase, induction in yeast, 60–61
actin gene introus in yeast, 46
activation (metabolic) factors affecting assays, 191–192
activation, metabolic, use of control compound in tests, 180
activation of mutagens (carcinogens) by yeast test systems containing endogeneous cytochrome P-448/P-450, 189–191
activation of mutagens in micro-organisms, 189
activation of mutagens, in yeast, 130–157, 188–195
ADE genes, red pigments of, 170, 174, 176
ade_1 (red) yeast mutant, 18–20
ADE_2 genes, in yeast test systems, 178
Adriamycin, petite colony formation in yeast by, 26–29
Adriamycin, toxicity of, in yeast, 26–29
aeration, in cytochromes P-450 production in yeast, 136–138
aeration, optimal, in sporulation of yeast, 187
α-factor in yeast, 61–66
a-factor in yeast, 61–66
α-galoctosidase, induction by galactose in yeast, 58
agar media (top and bottom) in test systems, 182
alcohol dehydrogenase removal, in yeast test system methodology, 185
allyl alcohol resistance, and forward (point) mutations in yeast, 185
alkylating agents, in yeast test system methodology (washing with sodium thiosulphate), 182
Ames test, large-scale trials of, 194
Ames test, methodology with, 182
Ames test, outline of, 169
Ames test, problems with using the S9 liver fraction as activator, 191
ampicillin resistant (Amp^r) genes, on yeast plasmids, 49
analysis and interpretation of tests in yeast, 192–195
analysis (statistical) of yeast test system data, 192
analysis of gene relationships, in yeast, 12–40
aneuploidy in yeast, 177–179
aneuploidy, mitotic and meiotic in yeast, 177–179
anaeuploidy scoring, in yeast disomic spores, 187
animals, mitochondrial DNA as carcinogen target, 24
antibiotics and bacterial growth, 31
antibiotic resistance, inheritance in yeast, 32–34
antibiotics, tested in yeast, 29–32
antibody production, by yeast, 42
antibodies to cytochromes P-450, use of, 126
application for high copy number plasmids and high expansion yeast promoters, 195
applications of yeast in biotechnology, 42
arginine permease removal, in yeast, test system methodology, 185
Aroclor-treated rats, in preparation of liver s9 faction, 189
ARS: autonomoulsy replicating sequence, 41
ARS plasmids in yeast, 49–50
ascosopores, in yeast, 14–15
ascus of yeast, 14
assay of cytochrome P-448 in yeast, 153, 156
assay of genotoxicity of chemicals, using yeast, 10
assay (test) systems, for mutagens and carcinogens in yeast containing endogeneous cytochrome P-448/P-450, 189–191
assays for mutotic gene conversion, in yeast, 181
assays (in yeast) constructed by recombinant DNA methods, 195
AUG in iniation codon, in ORF, 44
auxotrophic colonies, in yeast, 14

Bacillus megaterium, cytochromes, P-450 from, 126–127
bacteria, for detection of mutagens (and carcinogens), 9
bacterial growth, effect of tetracycline, antibiotics, 31
bacterial tests, spot-tests and fluctuation tests, 182
bay-region of polycyclic hydrocarbons, 124
benzo(a)pyrene, activation by yeast test system containing endogeneous cytochrome P-448/P-450, 190
benzo(a)pyrene, activation of, 124
benzo(a)pyrene and cytochrome P-450, oxygen insertion, 188–189
benzo(a)pyrene, as control chemical in yeast test systems, 180
benzo(a)pyrene, as inducer of cytochrome, P-448, in *Sarcharomyces cerevisiae*, 148–151
benzo(a)pyrene hydroxylase activity, of yeast cytochrome P-448, 143–148
benzo(a)hydroxylase, fluorimetric assay for, 156
benzo(a)pyrene, hydroxylation in yeast, 143–148
benzo(a)pyrene metabolism, by yeast enzymes, 143–148
β-galactosidase in meauring induced DNA damage, 195
binding studies, of promutagens and putative carcinogens to yeast cytochrome P-448, 151–152
biochemical markers, in yeast, 14
biosynthesis of cytochromes P-450, in yeasts, 131–139
biosynthesis of haem in yeast, 59–60
biotechnology, applications of yeast, 42
biotransformations of chemicals, and problem of metabolism in genotoxicity, 188–192
bp: basepair, 41
breaks in DNA, double stranded, use of with integrative vectors, 51–52
budding in yeast tests, 182
budding of yeast, coincidence with DNA replication, 183

canavanine resistance, 179
canavanine resistance, and forward (point) mutations in yeast, 185
Candida species, occurrence of cytochromes P-450 in, 130–131
Candida tropicalis, occurrence of cytochromes P-450 in, 130–131
capping (5′), by guanine nucleotides, 44
capping (3′), by polyadenylation, of yeast messengers BNA, 43
carcinogen (mutagen) activation, comparison of *in vivo* and *in vitro* products, 191
carcinogen (mutagen) activation by indogenous cytochrome P-448/P-450, in yeast test, systems, 189–191
carcinogen (mutagen) metabolism, for genotoxicity, problems of, 188–192
carcinogen-target, mitochondrial DNA in, animals, 24
carcinogen testing, advantages of using yeast, 169–170
carcinogens, activated by cytochromes P-450, 121–125
carcinogens and mutagens, factors affecting metabolic activation of, 191–192
carcinogens, attack on yeast mitochondria, 172
carcinogens (putative) binding studies to cytochrome P-448 in yeast, 151–152
carcinogens, damage unrelated to genotoxicity, 177
carcinogens, effect on petite colony formation in yeast, 23–29
carcinogenesis, xenobiotic transformation prior to, 168–169
carcinogenicity and mutagenicity tests in yeast, analysis and interpretation of, 192–195
carcinogenicity, dependence on mutagenicity, 168–169
carcinogenicity, of benzo(a)pyrene, 124
carcinogenicity testing, methodology details, 10
carcinogenicity testing — see mutagenicity testing by mutotic recombination in yeast in UK guidelines, 193
cells, mammalian, and protein half-lives, 42
CEN-containing plasmids, one per yeast cell, 50
centrifugal elutrication, of yeast, 186
centrifugal separation of yeast cells in exponential phase of growth, 186
centromere (CEN)-containing plasmids, one per yeast cell, 50
centromere, of yeast, 14
chemical damage to DNA, and repair in yeast, 92–95
chemicals, mutagenicity of, see EEC directive 83/5701 1984, 193
chemicals transformed to mutagens, in yeast test systems containing endogeneous cytochrome P-448/P-450, 189–191
chlorimipramine, effects on growth of yeast, 21
chromatids (bivalents), of yeast, 14
chromosome 21, in Down's syndrome, 177
chromosome VII, monosomy of, 178
chromosomes, yeast, 174–178
circu 2+, the endogenous 2 mucron (μ or μm) plasmid of yeast, 50

circular DNA, of yeast mitochondrion, 22
CIS-acting elements, in yeast, 47
cloned cytochrome P-450 in yeast, 195
clones of yeast, toxicity of chemicals in, 181
coding strand, of DNA, 42, 43
codons of mitochondrial DNA, 22
colony type, and genetic change in yeast, 180–181
continuous culture, for production of cytochromes P-450 in yeast, 135–136
control considerations, in yeast test systems, 193
control levels, if high in diploids, 181
controls for solvent used, in yeast test systems, 180
controls in yeast test systems, 192–195
cost-effectiveness, of short term tests, 10
Coulter counter, use with yeast test cultures, 183
crossing-over, between yeast chromatids, 14
cumene hydroperoxide, to drive cytochrome P-450 systems, 118, 119
cytochrome b_5, relationship with cytochrome P-450, 117
cytochrome P-448, activity measurement of, 125
cytochrome P-448, assay of, in yeast, 153, 156
cytochrome P-448, binding studies of promutagens and putative carcinogens to, 151–152
cytochrome P-448 from *Saccharomyces cerevisiae*, benzo(a)pyrene hydroxylase activity of, 143–148
cytochrome P-448 in yeast, induction by benzo(a)pyrene of, 148–151
cytochrome P-448 in yeast, see all cytochrome P-450 references, e.g., 130–157
cytochrome P-448, purification of from *Sacchromyces cerevisiae*, 154–156
cytochrome P-448, role inside yeast cells, 9
cytochrome P-448/P-450 from rat liver, use of in test systems, 9
cytochrome P-450, antibodies to, 126
cytochrome P-450, and benzo(a)pyrene, oxygen insertion, 188–189
cytochrome P-450_{cam}, 125–127
cytochrome P-450, control by cyclic AMP in yeast, 134
cytochrome P-450, improvement by protein engineering or protein modification, 195
cytochrome P-450, induction in mammals, 121–125
cytochrome P-450, equation for reaction of, 115
cytochrome P-450 (mammalian), expression in yeast, 195
cytochrome P-450, redox potential of, 145
cytochrome P-450, from *Pseudomonas putida*, 125–126
cytochrome P-450 from yeast, general comments on, 190
cytochrome P-450, multiple forms of, 121–125
cytochrome P-450, optimization in yeast of, 134
cytochrome P-450, reaction mechanism of, 116–121
cytochrome P-450, structure and role, 116–121
cytochrome P-450 system, in yeast, discovery of, 170
cytochrome P-450 systems, second electron from cytochrome, b_5, 117
cytochrome P-448/P-450, (endogenous) in yeast test systems in yeast given in 20% glucose, 189–191
cytochrome P-450 versus deactivating enzymes, 194
cytochrome in yeast, general discussion, 189–190
cytochromes P-450, and mutagenesis, 124
cytochromes P-450, biosynthesis of, in yeasts, 131–139
cytochromes P-450, effect of oxygen and aeration in yeast, 136–138
cytochromes P-450, from *Bacillus megaterium*, 126–127
cytochromes P-450, from eukaryotic micro-organisms, 128–130
cytochromes P-450, gene families with, 123
cytochromes P-450, gene sequences of, 122
cytochromes P-450, in activation of carcinogens, 121–125
cytochromes P-450, in alkane metabolism in yeast, 130–131
cytochromes P-450, in *Saccharomyces cerevisiae*, 131–157
cytochromes P-450, in yeast, induction by ethanol, 137–139
cytochromes P-450 (endogenous), in yeast test systems, 189–191
cytochromes P-450, occurrence in *Candida species*, 130–131
cytochromes P-450, occurrence in yeasts, 130–139
cytochromes P-450, of mitochrondria, 126
cytochromes P-450, of prokaryotic micro-organisms, 125–128
cytochromes P-450, production, in continuous culture in yeast, 135–136
cytochromes P-450, receptors in induction of, 123
cytochromes P-450, specificity of, 122
CYC1 genes in yeast, 59–60
cyclic AMP, in control of yeast cytochrome P-450, 134
cycloheximide, as inhibitor of protein biosynthesis, 21
cycloheximide, sensitive and resistant yeast, 175, 178
cyclophosphamide, activation by yeast test system containing endogeneous

cytochrome P-448/P-450, 190
cyclophosphamide, as control chemical in yeast test systems, 180

D4, yeast, 174
D4 and D6 yeast, in test systems containing endogeneous cytochromes P-440/P-450, 189–191
D4, D5, D7 and JD1 yeasts, validation of, 194
D6 yeast, in detection of nonmutagenic species, 194
D7, yeast, 174–176
D7 yeast, application of in mutagenicity testing, 174
D7 yeast, sporulation conditions for, 186
damage inducible (DIN) genes in yeast, 195
damage to DNA, and gene induction, 85–86
damage to DNA, by chemicals, repair on yeast, 92–95
data, quantitative and qualitative interpretation of, in yeast test systems, 182
deactivating enzymes versus cytochrome P-450, 194
demethylation in yeast, of lanosterol, 139–143
diethylstilboestrol, aneuploidy assays in yeast, 194
dimethylhydrazine, with yeast test systems, 180
dimethylsulphoxide, as solvent in yeast test systems, 180
DIN genes in yeast, 195
diploid yeast, high control level with, 181
diploid yeast, sporulation check on, 180, 181
diploid zygotes, in yeast, 12, 13
DIS 13, in yeast, 178
disomics (n+1), selection for in yeast, 178
DMSO, dimethylsulphoxide, solvent used in yeast test systems, 180
DNA, coding strand of, 42, 43, 44
DNA damage by chemicals, and repair in yeast, 92–95
DNA damage, by reactive oxygen species, 194
DNA damage, inducible, assessed by measuring β-galactosidase in yeast, 195
DNA digest, PAGE photograph of result, 36
DNA, double-stranded breaks in, use of with integrative vectors, 51–52
DNA, entry into yeast sphaeroplasts using lithium ions, 48
DNA, entry into yeast sphaeroplasts using polyethylene glycol and calcium ions, 48
DNA, excision repair of, 79–85
DNA, flanking the ORF, 42, 43
DNA footprinting in yeast, 58
DNA (exogenous), in yeast transformation, 47–49
DNA, integrative vectors in yeast, 51
DNA introduction into yeast cells, 48–49
DNA mutation in mitochondria, 23–29
DNA, of mitochondria, 22
DNA of mitochondrion, as a site of oncogenesis, 39
DNA mitochondria, effect of Adriamycin, 26–29
DNA of mitochondria, in animals, as target for carcinogens, 24
DNA of yeast mitochondria, cleavage by restriction endonucleases, 36, 37
DNA, recombinant, and RAD6 repair pathway, 89
DNA, recombinant and the RAD52 gene, 91–92
DNA, recombinant for construction of new (yeast) assays, 195
DNA repair and mutagenesis, 172–173
DNA repair, and mutagenesis study, 84
DNA repair and problem of metabolism in genotoxicity, 188–192
DNA repair, and RAD mutants of yeast, 78–85
DNA repair, and radiation, 78–79
DNA repair in yeast, after chemical damage, 92–95
DNA repair pathways, general consideration of, 74–77
DNA repair, proteins involved in, 83
DNA replication coincidence with yeast budding, 183
DNA transformations, in yeast, 41
dominant genes, in yeast, 14
dose level, of test compounds in yeast, 180
dose of exposure, and mutant yield calculation, 192
double stranded breaks in DNA, use of with integrative reactors, 51–52
Down's syndrome, 177
downstream DNA, to 3′ side of ORF, 42, 43
Drosophila ADE8 gene, in yeast, 46
drug metabolism in yeast, effect of phenobarbitone, 28
drug metabolism, use of liver S9 fraction *in vitro*, 194
drug-resistance mutations, in yeast, 20–22
drug resistance, in yeast, 32–34
drug screening, in yeast, 12–40

E. coli, DNA repair in, 74–77
economics of short-term tests, 10
EEC directive 83/5701 1984, four recommended tests for mutagenicity of chemicals, 193
EEC legislation on pharmaceuticals, and short term tests, 193
efficiency of promoters, 44
endogenous cytochrome P-448/P-450 in

yeast test systems, 189–191
endonucleases (restriction), action on yeast mitochondrial DNA, 36, 37
endpoints, genetic, of test systems, 170–179
engineering of proteins, see protein engineering, 195
enzyme-activated mutagens and carcinogens, in yeast test systems containing endogenous cytrochrome P-448/P-450, 189–191
enzyme expressions level in yeast, use of lac Z fusion to determine it, 51–52
enzymes from yeast, in benzo(a)pyrene metabolism, 143–148
enzyme induction, in yeast, 56–61
enzyme-mediated biotransformation of chemicals in genotoxicity, 188–192
enzymes (deactivating) versus cytochrome P-450, 194
enzymology, importance of, in yeast, 10
epistasis groups, in yeast genes, 87, 89–92
episome-like inheritance of tetracycline resistance in yeast, 34
episome-like inheritance of TC resistance, 34
epoxide hydratase, use of inhibitors of in test systems, 191
ergosterol, pathway of biosynthesis in yeast, 139–143
erythormycin resistance, in yeast, 32–34
Escherichia coli, SOS repair system of, 195
ethanol, in cytochromes P-450 production in yeast, 137–139
ethoxyresorufin, as specific substrate for cytochrome P-448, 125
ethyl methane–sulphonate, as mutagen in yeast, 180
eukaryotes, higher, 12
eukaryotic micro-organisms, cytochromes P-450 from, 128–130
evaluation of toxicity, from yeast test systems, 192–195
excision repair of DNA, 79–85
experimental procedures, for tests, 179–188
experimental protocols, with yeast test systems, 181–188
expansion level in yeast, use of lac fusion, 51–52
exposure dose and mutant yield, 192

flanking DNA (i.e. not ORF), 42, 43
flutuation tests, with yeast, 182
forward mutation (point) in scoring genotoxicity, 184–185
frame shift, absence at gene function, 52
frameshift mutations, in yeast test systems, 185
free radicals, in yeast test systems, 26–29

GAL genes in yeast, induction of, 57–58
GAL2 gene, in yeast, 174
galactose, as inducer of GAL genes in yeast, 57–58
galactose induction of MEL1 gene in yeast, 58
GAL1 and GAL10 genes in yeasts, 55
gene conversion assays, in yeast, 181
gene conversion endpoints, in methods, trials in yeast assessed, 194
gene conversion (mitotic) scoring of, 183
genes, damage inducible in yeast, 195
gene, definition of, 42
genes, epistasis groups, in yeast, 87, 89–92
gene families, of cytochromes, P-450, 123
gene for actin, in yeast, intronus in, 46
genes, foreign, expression of in yeast, 9
gene function, absence of frame shift, 52
gene fusion functions, reading frame in register, 52
gene fusions, in yeast, 47, 48
genes (DIN) in yeast, 195
genes (MAT) in yeast, 61–66
genes in yeast for iso-cytochrome C production (CYC1), 59–60
gene in yeast, TRP5, 174
gene induction by DNA damage, 85–86.
gene induction: mechanism of, 41–72
gene introns, in action and MAT α1 genes in yeast, 46
gene linkage, in yeast, 14
gene recombination, in yeast, 14
gene (RAD) regulation in yeast, 195
genetic regulation, of cytochrome P-450 in yeast, 135–136
gene repression, mechanism of, 41–72
gene sequences, of cytochromes P-450, 122
genes, termination signals in, 44
gene, TRPI, in yeast, 49
genes of yeast, protein repressors of, 55–56
genes of yeast, RNA2–RNA11, 47
gene, yeast, GAL2, 174
gene, yeast, LEU1, 174
genetic analysis, in yeast, 12–40
genetic analysis in yeast, of extragenic control-modifying mutations, 47
genetic change, and yeast colony type, 180–181
genetic diseases, Down's syndrome, 177
genetic endpoints, in test systems, 170–179
genetic engineering, in yeast, 9, 12
genetic stability, in yeast test systems, 180–181
genetic studies, and the cell cycle in yeast, 102–105
genetic studies on mutagenesis in yeast, 102–105
genetics of yeast, 12–40
genetics, yeast mitochondrial, 22–40
genotoxic agents, assessment of mutagenic

efficiency, 193
genotoxic assay, in gene conversion endpoint trials, assessment in yeast, 194
genotoxic species, generated by yeast test systems containing endogenous cytochrome P-448/P-450, 189–191
genotoxicity, and other possible effects of carcinogens, 177
genotoxicity, requirement for metabolism of carcinogen (mutagen), 188–192
genotoxicity, scoring of, 184–188
glucose (20%), in optimization of carcinogen activation in yeast test system containing endogenous cytochrome P-448/P-450, 189–191
glucose repression in yeast, 57–58
glucose repression of MAL genes, in yeast, 58–59
glutathione peroxidase, protective role in yeast, 27
glycosylation of proteins, in yeast, 42
growth of bacteria, effect of tetracycline antibiotics, 31
growth of yeast, effect of chlorimipramine, 21
growth of yeast, effect of tetracycline antibiotics, 29–32
guidelines, for safe handling of test compounds, 179
guidelines, UK and EEC, for mutagenicity testing, 193–194

haem biosynthesis, in yeast, 59–60
haem, induction of CYC1 genes in yeast, 59–60
haemocytometer, use with yeast test cultures, 183
half-life of proteins, in circulation, 42
handling, safety, of test compounds, 179
haploid yeasts, 169
hazard evaluation, using yeast test systems, 193–195
HB_sAg, production in yeast, 42
HB_sAg, virus, particles produced in yeast, 42
heat shock, activation of PGK promoters by, in yeast, 53
heat shock in yeast, effect on protein biosynthesis, 66–68
heat shock, selective translation of certain mesenger RNAs, 44
hepatitis B virus, production of surface antigen of, in yeast, 42
heteroallelic mutagens, definition of, 15–16
heterologous proteins, expressed in yeast, 42
high copy number plasmids, uses in yeast, 46, 47
MIS4 gene, of yeast, 175
hormonal control of mammalian cells, and splicing, of messenger, RNA, 47
host-mediated test systems, 191
hybrid promoters, in yeast, 47, 48
hybrid proteins, in yeast, value of, 52
hydrazine (dimethyl) with yeast test systems, 180
hydroxyurea, with yeast test systems, 180

incubation of plates, time of, in yeast test systems, 183
induced enzyme-activated mutagens and carcinogens, in yeast test systems containing endogenous cytochrome P-448/P-450, 189–191
induced prototrophs, selection of, 181
inducers of cytochrome P-448, in *Saccharomyces cerevisiae*, 143–151
inducible genes in yeast, six damage-inducible, 195
inducibility of mutagenesis, 99–101
induction by benzo(a)pyrene, of cytochrome P-448 in yeast, 148–151
induction by ethanol, of cytochromes P-450 in yeast, 137–139
induction by galactose of GAL genes in yeast, 57–58
induction of acid phosphatase in yeast, 60–61
induction of CYC1 genes by haem, in yeast, 59–60
induction of cytochrome P-450, in mammals, 121–125
induction of cytochromes P-450, receptors in, 123
induction of enzymes in yeast, 56–61
induction of GAL genes in yeast, 57–58
induction of genes, by DNA damage, 85–86
induction of genes, in yeast, 41–72
induction of genetic endpoint assessment, methodology of, 182
induction of maltase in yeast, 58–59
induction of maltose permease, in yeast, 59
induction of mating type genes in yeast, 61–66
induction of mutant yield, calculation for yeast test systems, 192
induction of reverse mutation, in homoallelic yeast, 181
inheritance, episome-like of TC resistance, 34
inheritance in yeast, of tetracycline resistance, 34
inheritance of drug resistance, in yeast, 32–34
inhibitors of epoxide hydratase, use in test systems, 191
inheritance of petite mutant of yeast, 32–34
inhibitors of mitochondrial protein

biosynthesis, 22
inhibition of protein biosynethesis, by cycloheximide, 21
inhibitors of protein biosynthesis in yeast, tetracycline, antibiotics, 29–32
initiation of transcription, and TATA box, 42
immunoglobulin (antibody, production of, by yeast, 42)
integrative vectors, for yeast, 49–50
interaction of nucleus and mitochondrion, in yeast, 35–38
interpretation and analysis, of tests in yeast, 192–195
introns in actin gene, in yeast, 46
introns, in lariat form, in yeast, 46
introns in MAT α1 gene, in yeast, 46
introns in *Neurospora*; 46
introns in *Schizosaccharomyces pombe*, 47
introns in yeast, 42
introns in yeast, 44
introns in yeast messenger RNA, seven nucleotide sequence required for processing of, 46
introns, splicing out of, in yeast, messenger RNA, 46
ISO-cytochromes C (CYC1) genes in yeast, 59–60
isoenzymes of cytochrome, P-450, 121–125
invertase, SUC2 gene, in yeast, 57
in vitro, studies on drug metabolism using S9 liver fraction, 194
in vitro, transcription in yeast, 41
in vitro test systems, general aspects, 189
in vivo and *in vitro*, comparison of mutagen (carcinogen) activation in, 191

Kd: kilodaltons, 41

lac Z (β-galactosidase) gene fusion, use of to determine level of expression in yeast, 51–52
ianosterol, demethylation in yeast, 139–143
lariat-introns, in yeast, 46
leader regions, of yeast messenger, RNA, 44
legislation and guidelines — see UK guidelines with mitotic recombination test in yeast, for mutagenicity, 193
legislation, EEC, on pharmaceuticals, 193
lethal effect, measurement of in test systems, 182
LEU1 gene, in yeast, 174
LEU2 gene, as selectable marker for transformation in yeast, 48
level of dose, of test compounds, in yeast, 180
life cycle of yeast, 12–17
linkage of genes, in yeast, 14
lithium ions, in transforming, yeast with DNA, 48
liver S9 fraction, application to general *in vitro* studies of drug metabolism, 194
liver S9 fraction, from Aroclor-tested rats, 189
liver S9 fraction, preparation of, and its use, 188, 189
liver supernatant fraction (S9), 169
logarithmic phase cell, in yeast test systems, 183–184

magdala red, to identify biochemical mutants in yeast, 14
MAL (maltase) genes in yeasts, 58–59
maltose, as inducer of maltase in yeast, 58–59
maltose permease, induction in yeast, 59
maltose regulation, in yeast, 58–59
mammalian cells and protein half-life, 42
mammalian cells, correct splicing of yeast genes transcripts in, 47
mammalian cells, hormonal control and splicing of messenger RNA in, 47
mammalian cytochrome P-450, expressed in yeast, 195
markers, biochemical, in yeast, 14
master-culture of test yeasts, 181
MAT α1 gene, introns in, in yeast, 46
MAT α1, protein factors, effect on gene transcription, in yeast, 55
MAT genes in yeast, 61–66
mating factors, α, in yeast, 12
mating factor, a, in yeast, 12
mating, illegitimate, in yeast, 179
mating type genes, in yeast, 61–66
mating-type, switch events in, 179
mechanism of reaction, cytochrome P-450, 117
media recipes, for yeast test systems, reference for, 182
meiotic cell cycle, and yeast test systems, 186
meiotic geast test systems (sporulation), 187
meisois in yeast, 14
MEL1 gene in yeast, induction by galactose, 58
messenger RNA in yeast, all have poly A tails, 44
messenger RNA, of yeast, leader regions in, 44
messenger RNA of yeast, no obvious Shine-Dalgarno sequence in, 44
messenger RNA, polyadenylation of, 43
metabolic activation, factors affecting this in test systems, 191–192
metabolic activation of mutagens, in yeast, 130–157, 188–195
metabolic activation, use of control compound in tests, 180

metabolisms in genotoicity, problem of, 188–192
metabolism of drugs, use of S9 fraction as model system *in vitro*, 194
methodology, for tests, 179–188
methodology for yeast test systems, with endogenous cytochrome P-448/P-450 enzymes, 189–191
methodology in tests, practical requirements of use of yeast, 10
methodology, of induction of genetic endpoint assessment, 182
methodology of yeast test systems, "treat-and plate method", 182
methodology, using S9 fraction, in yeast test systems, 189
3-methylcholanthrene-induced cytochrome P-450, 121–125
micro-organisms and activation of mutagens, 189
mitochondria, attack by chemical carcinogens, 172
mitochondria, inhibitors of protein biosynthesis in, 22–23
mitochondria in yeast, and tetracycline antibiotics, 29–32
mitochondria of yeast, 22–40
mitochondria, protein biosynthesis in, 22–23
mitochondria, pyrimidine dimer repair in, 81
mitochondrial cytrochromes P-450, 126
mitochondrial DNA, 22
mitochondrial DNA, as a site of oncogenesis, 39
mitochondrial DNA, effect of Adriamycin, 26–29
mitochondrial DNA, in animals, as target for carcinogens, 24
mitochondrial DNA mutation, 23–29
mitochondrial drug resistance, inheritance in yeast, 32–34
mitochondrial genetics, in yeast, 22–40.
mitochondrial-nuclear interactions, in yeast, 35–38
mitochondrial petite mutation, in yeasts, 23–38
mitochondrial recombination, in yeast, 36, 37
mitotic cell cycle, and yeast test systems, 186
mitotic gene conversion assays, in yeast, 181
mitotic crossing-over, in scoring genotoxicity (red and pink colours on limited adenine), 184
mitotic gene conversion, in scoring genotoxicity, 184
mitotic gene conversion, scoring of, 183
mitotic recombination, in yeast, 17–20
mitotic recombination yeast test, in UK guidelines for mutagenicity testing, 193
MNNG, in yeast, test systems, 180
MNNG, preferential action at the replication fork in DNA replication, 183
molecular biology, importance of, in yeast, 10
money value, of short term test, 10
monosomy of chromosome VII, 178
multiple forms of cytochrome P-448, induction in yeast, 149
multiple forms, of cytochrome P-450, 121–125
mutagens (indirect-acting) activated by endogenous cytochrome P-448/P-450 in yeast, 189–191
mutagen (carcinogen) activation by endogenous cytochrome P-448/P-450, in yeast test systems, 189–191
mutagen activation by *Saccharomyces cerevisiae*, of a variety of mutagens, 147–148
mutagen (carcinogen) activation, comparison of *in vivo* and *in vitro*, 191
mutagen activation in micro-organisms, 189
mutagen activation, in yeast, 130–157, 188–195
mutagens and carcinogens, factors affecting metabolic activation of, 191–192
mutagen efficiency of genotoxic agents, assessment of, 193
mutagen in yeast, ethyl methane-sulphonate, 180
mutation in yeast, spontaneous and induction reversion, rate of, 180–181
mutagen (carcinogen) metabolism, for genotoxicity, problems of, 188–192
mutation, reverse, scoring of, 183
mutation scoring, 183
mutagen testing, advantages of using yeast, 169–170
mutagen treatment, throughout sporulation of yeast, 187–188
mutagenesis and cell cycle in yeast, 102–105
mutagenesis and DNA repair, 172–173
mutagenesis and UV radiation, studies on micro-organisms, 172–173
mutagenesis by X-rays, 87
mutagenesis in yeast, and cell cycle studies, 102–105
mutagenesis in yeast, general considerations, 95–106
mutagenesis, inducibility of, 99–101
mutagenesis repair, the RAD6, pathway for, 86–89
mutagenesis, specificity of, 101–105
mutagenesis study, and DNA repair, 84
mutagenesis, timing of in yeast cells, 104–105
mutagenesis, with cytochromes P-450 products, 124
mutagenic result, biological significance threshold of, 193
mutagenic species, production, and problems of metabolism of genotoxicity, 188–192
mutagenic yield curve, integral from, 193

mutagenicity and carcinogenicity tests in yeast, analysis and interpretation of, 192–195
mutagenicity, and other possible effects of carcinogens, 177
mutagenicity and S9 MIX, 124
mutagenicity, of benzo(a)pyrene, 124
mutagenicity of chemicals, see EEC directive 83/5701 1984, 193
mutagenicity requirement, for carcinogenicity, 168–169
mutagenicity testing, methodology details, 10
mutagenicity testing, mitotic recombination yeast test in UK guidelines, 193
mutagenicity testing UKEMS guidelines, 193
mutagenicity test, using *Salmonella* (Ames test), 169
mutagenicity tests, using yeast D7, 174
mutant, ADE pathway, and red pigment, 170, 174
mutants of yeast, use of TRP1 gene, 51–52
mutants, yeast, conditional lethal (temperature-sensitive) in, 47
mutants (viable) production of, assessment method, 193
mutant (viable) yield, graphical method and calculation, 192
mutant yield induction, calculation for yeast test systems, 192
mutations analysis in yeast using integrative vectors, 51
mutations, frameshift in yeast test systems, 185
mutations, for drug resistance, 20–22
mutations, forward, 170–171
mutations, heteroallelic, in yeast, 15–16
mutation (point) in scoring genotoxicity, 184–185
mutations (point) in yeast, and canavanine and allyl alcohol resistance, 185
mutations, of extragenic control-modifying type, 47
mutations, point, 170–172
mutations, reverse, 171–172
mutation, spontaneous in yeast, 97–99

n-alkanes, metabolism in yeasts by cytochromes P-450, 130–131
N-demethylase activity in yeasts, 131
non-disfunction, leading to monosomy of chromosome VII, 178
Neurospora introns, 46
NIM shift, during hydroxylation reactions, 129
nitrite, with yeast test systems, 180
N-methyl-N-nitroso-guanidine (MNNG), in yeast test systems, 180
nonmutagenic species, detected in the yeast D7, 194
Northern blot, for analysis of RNA, 51
nuclear-mitochondrial interactions in yeast, 35–38

oncogenesis theory, 39
open-promoter complex, 42, 43
optimization of cytochrome P-450 in yeast, 134
ORF, as protein coding region of genes, 42, 43, 44
ORF, A+T rich in yeast genes, 44
ORF, containing activator sequence in PGK gene of yeast, 55
ORF, downstream DNA to 3′ side of, 42, 43, 44
ORF, initiation codon AUG, 44
ORF, open reading frame, 41
ORF, upstream DNA to 5′ side of, 42, 43, 44
origin of replication (OR1), yeast plasmids, 49
overview of yeast test systems, 193–195
oxygen, in cytochromes P-450 production in yeast, 136–138
oxygen (reactive) species and DNA damage, 194

PAGE of yeast mitochondrial DNA digests, 36
PAGE photograph, of DNA digest of yeast mitochondrial DNA, 36
petite colony formation by Adriamycin, 26–29
petites, increased sensitivity to chemicals in, 181
petite mutant, inheritance in yeast, 32–34
petite mutant, JD1, of yeast, 175
"petites", scoring of, 185
PGK glue in yeast, ORF activator sequence of, 55
PGK (phospho-glycenekinase) promoters, in yeast, 53
pharmaceuticals, EEC legislation on, 193
pharmacotinetics of xenobiotics, 195
phenobarbitone, effects on drug metabolism by yeast, 28
phenobarbitone-induced cytochrome P-450, 121–125
PHO genes in yeast, 60–61
phosphatases induction in yeast, 60–61
phosphoglycero-kinase (PGK) promoters in yeast, activation of, by heat shock, 53
photoreactivation in DNA repair, 77–78
pink colours of yeast, 175
plasmids (high copy number) application for, 195
plasmids containing CEN; one per yeast

cell, 50
plasmid endogenous (2μ:circu2+) in yeast, 50
plasmids, for yeast use, 49–51
plasmids (yeast) having Amp^r and Tet^r genes, 49
plasmid (2μ) in yeast, 49–51
plasmids (ARS) in yeast, 49–50
plasmid (YRPT) in yeast, 49
plasmid stability, in yeast, 49, 50
plasmid uptake, by yeast sphaeroplasts, 48
plasmids with high copy number, uses in yeast, 46, 47
plasminogen (tissue) activator, production in yeast, 42
point mutations, 170–172
point mutation, in scoring genotoxicity, 184–185
poly A — enriched messenger RNA, use of, 51
poly A, short tails in yeast messenger RNA, 44
poly A tails, in all yeast messenger RNA, 44
polyacrylamide gel electrophoresis, of mitochondrial DNA digest, 36
polyadenylation, of messenger RNA, 43
polycyclic hydrocarbons, bay-region of, 124
polyethylene glycol, use in transforming yeasts with DNA, 48
polyploid zygotes, 178
procedures, experimental, for tests, 179–188
processing and splicing of yeast introns, seven nucleotide sequence required for, 46
prokaryotic forms of cytochromes P-450, 125–128
promoters (of tumours), 177
promoters, hybrid constructs in yeast, use of, 47, 48
promoter efficiency, 44
promoters (high expression) applications in yeast, 195
promoter fusions, in yeast, 47, 48
promoter of transcription, 42, 43, 44
promoters of tumours, and yeast detection systems, 194
promoter, open complex, 42, 43
promutagens, binding studies to cytochrome P-448 in yeast, 151–152
promutagens, lipophilic nature at physiological pH of, 188
protein biosynthesis, effects of heat shock on, in yeast, 66–68
protein biosynthesis in mitochondria, inhibitors of, 22
protein biosynthesis inhibition, by cycloheximide, 21
protein biosynthesis inhibitors, antibiotics tested in yeast, 29–32
protein biosynthesis machinery, switch-on in inductions, 10
protein biosynthesis, within mitochondria, 22–23
protein coding region of genes, as ORF, 42, 43
protein engineering, for cytochrome P-450, application of, 195
protein gene expression, 42
protein glycosylation, in yeast, 42
proteins, half-life in circulation, 42
protein, heterologous, expression of in yeast, 42
protein hybrids in yeast, value of, 52
protein modification for cytochrome P-450, application of, 195
protein overproduction, in yeast, 50
protein synthesis, scheme for production of mature messenger RNA, 43
protein tailoring, *see* protein modification, 195
protocols for yeast test systems, 181–188
protoplasts (sphaeroplasts) of yeast, preparation and use of, 48
prototrophic colonies, in yeast, 14
prototrophic colonies, of yeast, 179
prototrophs, induced, selection of, 181
prototrophs, spontaneous, selection problem, 181
Pseudomonas putida, cytochrome P-450 from, 125–126
pyrimidine dimer removal, in DNA repair, 80–81

RAD52 and recombination processes, 89–91
RAD52 gene, and recombinant DNA studies, 91–92
RAD gene regulation, 195
RAD mutants, in *Saccharomyces cerevisiae,* 78–85
RAD6 pathway, for repair of mutagenesis, 86–89
RAD6 repair pathway and recombinant DNA studies, 89
radiation and DNA repair, 78–79
reading frame, in register across gene fusion functions, 52
receptors, in induction of cytochromes P-450, 123
recessive genes, in yeast, 14
recipes of media, for yeast test systems, reference for, 182
recombinant DNA, and RAD6 repair pathway, 89
recombinant DNA, and the RAD52 gene, 91–92
recombinant DNA, for construction of new (yeast) assays, 195
recombinant DNA technology, advent of, 9
recombination, in yeast mitochondria, 36–37
recombination, meiotic, in yeast (and test systems), 173–176
recombination, mitotic, in mutagenicity

testing in yeast (UK guidelines), 193
recombination, mitotic, in yeast (and test systems), 173–176
recombination of genes, in yeast, 14
recombination percentage formula, for yeast, 17
recombination processes and the RAD52 gene, 89–91
red adenine-deficient mutant in yeast cells (adei), 18
redox potential, of cytochrome P-450 from *Saccharomyces cerevisiae*, 145
repair-deficient strains of yeast, 172–173
repair of DNA, and mutagenesis, 172–173
repair of DNA, and problem of metabolism in genotoxicity, 188–192
repair of DNA, error-prone and error-free processes, 88
repair of DNA, general considerations, 74–77
repair of DNA, proteins involved in, 83
repair of mutagenesis, the RAD6 pathway, 86–89
replicate vectors for yeast, 49–50
replication fork, in DNA replication, 183
replication of DNA, coincidence with yeast budding, 183
replication origin (OR1) in yeast plasmids, 49
repression of genes, in yeast, 41–72
repressors of yeast genes, 55–56
resistance to canavanine in yeast, 179
resistance to drugs, in yeast, 32–34
resistance to erythromycin, in yeast, 32–34
resistance to tetracycline in yeast, inheritance of, 34
restruction endonucleases, used on DNA of yeast mitochondria, 36, 37
ribosomal proteins, accumulation in yeast, 47
RNA analysis, by Northern blotting, 51
RNA, enriched for Poly A, use of, 51
RNA2 — RNA11 genes of yeast, 47
RNA polymerase II, of yeast, 42–72
RNA splicing, in yeast, 42
reverse mutation induction, in homoallelic yeast cells, 181
reverse mutation (point) in scoring genotoxicity, 184–185
reverse mutation, scoring of, 183
reversion rates of yeast mutation, spontaneous and induced, 180–181
reversion, spontaneous in yeast, 174
revertant colonies per plate, problem of no increase in, 181
revertant colony frequency, per survivors, 181

S9 fraction, application to general *in vitro* studies of drug metabolism, 194
S9 fraction, from liver of Aroclor-treated rats, 189
S9 fraction, preparation from liver, and its use, 188, 189
S9 fractions, problems in use of, in all test systems, 191
S9 fraction, use with stationary-phase yeast cells, in tests, 182
S9 (9000 g liver supernant) fraction, 169
S9 mix and mutagenicity, 124
S9-mix (fraction) use in yeast test systems, methodology of, 189
Saccharomyces cerevisicae, advantages in test systems for mutagens (and carcinogens), 169–170
Saccharomyces cerevisiae, cytochromes P-450 in, 131–157
Saccharomyces cerevisiae, expression of heterologous protein in, 42
Saccharomyces cerevisiae, gene induction, 41–72
Saccharomyces cerevisiae, gene repression, 41–72
Saccharomyces cerevisiae, genetic effects and cell cycle in, 102–105
Saccharomyces cerevisiae, introns in, 44
Saccharomyces cerevisiae, mitotic cell cycle in, 186
Saccharomyces cerevisiae, mutagenens and the cell cycle in, 102–105
Saccharomyces cerevisiae, purification of cytochrome P-448 from, 154–156
Saccharomyces cerevisiae, RAD mutants, 78–85
Saccharomyces cerevisiae, RNA polymerase II of, 42–72
Saccharomyces lipolytica, alkane metabolism in, 131
Safe handling, of test compounds, 179
Salmonella (Ames) test, large-scale trials of, 194
Salmonella (Ames) test, problems with using the S9 liver fraction as activator, 191
Salmonella (bacterial) test methodology (Ames test), 182
Salmonella microsome test (Ames test), 169
Salmonella-microsome test *see Salmonella* (Ames) and Ames test, ???
Schizosaccharomyces pombe, in test systems, 170, 171
Schizosaccharomyces pombe, introns in genes, 47
Schizosaccharomyces pombe, mitotic cell cycle in, 186
Schizosaccharomyces pombe, splicing of introns in messenger RNA of, 47
scoring for aneuploidy, in yeast disomic spores, 187
scoring of genotoxicity, 184–188
scoring of mutations, 183
scoring "petites', 187

screening of drugs, in yeast, 12–40
segregation in yeast, 12–40
selective toxicity, in non-revertant yeast cells, 181
Shine-Dalgarno sequence, not obvious in yeast messenger RNA, 44
short poly A tails of yeast messenger RNA, 44
short-term tests, analysis and interpretation of, 192–195
short-term tests, and EEC legislation, 193
short-term tests, cost effectiveness of, 10
shuttle vectors, high stability and copy number of, 50
shuttle vectors, use in yeast, 49
signals for termination, in yeast genes, 44
solubility limits, of chemicals, in yeast test systems, 180
solvent controls, in yeast test systems, 180
solvents, used in yeast test systems, 180
SOS repair, in *E. coli,* 195
specificity, of cytochromes P-450, 122
specificity of mutagenesis, 101–105
sphaeroplasts of yeast, plasmid uptake by, 48
sphaeroplasts (protoplasts) of yeast, preparation and use of, 48
splicing in yeast, in control of ribosome formation, 47
splicing of intron transcripts, in *Schizosaccharomyces pombe*, 47
splicing, of RNA in yeast, 42
splicing (and processing) of yeast introns, seven nucleotide sequence required for, 46
splicing-out of introns, in yeast messenger RNA, 46
S. pombe, in test systems, 170, 171
spontaneous mutation, in yeast, 97–99
spontaneous prototrophs, problem of selection of, 181
spores (disomic) of yeast, 187
sporulation in yeast, conditions for, 186
sporulations in yeast, mutagen treatment during, 187–188
sporulation medium (SPM) for yeast, 14
sporulation of diploid yeast-test cultures, 180, 181
sporulation of yeast, good aeration requirement for, 187
spot tests, with yeast, 182
stability (genetic) in yeast test systems, 180–181
stability of plasmids in yeast, 49, 50
stationary-phase yeast cells, problems with, in tests, 181
statistical analysis, of yeast test system results, 192–195
storage in liquid nitrogen, of yeast-test master cultures, 181
stryrene activation by yeast test system containing endogenous cytochrome P-448/P-450, 190
SUC2 gene in yeast, 57
supernatant fraction (S9) from liver, use of in tests, 169
synchronized cultures of yeasts, use of, in test systems, 186

tailoring of proteins, *see* protein modification, 195
TATA box, and promoter modifying effects, 53–55
TATA box and UAS gene fusions, in yeast, 47
techniques of transformation of yeast, 48–49
test assays, requiring metabolic activation, factors affecting these, 191–192
test compounds, dose level for yeast, 180
test compounds, safe handling of, 179
tests, cost-effectiveness of, 10
tests for carcinogens, yeast petite colony formation, 23–29
test for mutagenicity, Ames test, 169
tests in yeast, analysis and interpretation of, 192–195
test methodology, 179–188
test systems, comparison of *in vivo* and *in vitro* conditions of activation, 191
test systems, factors affecting metabolic activation in, 191–192
test systems, genetic end points in, 170–179
test systems, host mediated assay, 191
test systems, trials of, 194
test systems (yeast) using endogenous cytochromes P-448/P-450, 189–191
test system using yeast, for tetracyclines, 29–32
test systems using yeast, free radicals in, 26–29
test systems, validation and hazard evaluation, 193–195
test systems, yeast, evaluation in a number of trails, 194
test systems, yeast, experimental protocols with, 181–188
test systems, yeast, methodology and use of S9 fraction from liver, 189
test systems, yeasts, solvents used in, 180
test systems, yeast, time of treatment with, 180
test systems, yeast, use of control chemical with, 180
test, two-sided, 192
tests, UK guidelines, 177
tests, yeast methodology in, 10
testing for mutagenicity, UKEMS guidelines, 193
testing for toxicity, using petite colony formation, 23–29
termination of transcription, site of, 44
termination signals, for transcription in

yeast genes, 44
tetrad analysis, in yeast, 17
tetrad formation, in yeast, 14
tetracycline antibiotics, and bacterial growth, 31
tetracycline resistance, inheritance of in yeast, 34
tetracycline-resistant (Tet^r) genes, on yeast plasmids, 49
tetracyclines, tested in yeast, 29–32
time of plates incubation, in yeast test systems, 183
time of treatment, in yeast test systems, 180
tissue plasminogen activator, production in yeast, 42
theory of oncogenesis, 39
threshold level for mutagenesis, in yeast, 180
toxicity evaluation, from yeast test systems, 192–195
toxicity of Adriamycin, 26–29
toxicity of chemicals, in normal and revertant clones of yeast, 181
toxicity, selective in non-revertant yeast cells, 181
toxicity testing, and petite colony formation, 23–29
trans-acting elements in yeast, 47
transcription initiation, and TATA box, 42
transcription *in vitro,* in yeast, 41
transcription promoter, 42, 43, 44
transcription termination site, 44
transcripts of Mammalian DNA, failure of splicing in yeast, 42
transcripts of yeast DNA, removal of introns from, 42
transformation in yeast, LEU2 glue as selectable marker for, 48
transformations, of DNA, in yeast, 41
translation of messenger RNA, effect of heat shock, 44
"treat- and plate" method, for assessment of toxcity and mutagenic effects, 182
treatment time, in yeast test systems, 180
trials of test systems, 194
trisomy of chromosome 21, Down's syndrome, 177
TRP1 gene, in yeast, 49
TRP5 gene in yeast, 174
TRP1, selectable gene use in yeast mutants, 51–52
TTC, in scoring of "petites", 185
tumour promoters, 177
tumour promoters, and yeast detection systems, 194
two-sided, yeast test method, 192
Ty element, negative regulation of, by a protein product in yeast, 56
Ty, transposable element in yeast, use of in transformation, 51
UAS, interaction with protein factors in yeast, 55
UAS: upstream activator sequence, 41
UK guidelines on mutagenicity testing, 193–194
UK guidelines, tests, 177
upstream DNA, to 5′ side of ORF, 42, 43, 44
uses of plasmids in yeast, at high copy number, 46, 47
UV radiation, and repair systems, 74–77, 80–81
UV radiation, on micro-organisms, and mutagenesis, 172–173
UV radiation, used as mutagenic agent, 14

validation of D4, D5, D7 and JD1 yeasts, 194
Value for money, of short term tests, 10
vector, multicopy integrative, uising Ty element in yeast, 51
vectors (replicative) for yeast, 49–50
vectors (integrature) for yeast, 50–51
vectors in yeast, high stability and copy number of shuttle vectors, 50
vectors, shuttle type, for yeast applications, 49
viability assessment, in yeast test systems, 182
virus particles, produced in yeast, 42

Xenobiotic metabolism by yeast, 143–148
Xenobiotic pharmacotinetus, 195
Xenobiotic transformation, prior to carcinogenesis, 168–169
X-rays, as mutagenicagent, 87

yeast, activation of mutagens and test systems for carcinogens, 130–157, 188–195
yeasts, advantages in test systems to mutagens (and carcinogens) 169–170
yeast a-factor, in mating, 61–66
yeast α-factor, in mating, 61–66
yeast, aneuploidy in, 177–179
yeast ARS plasmids, 49–50
yeast ascus, 14
yeast ascospores, 14–15
yeast, assay of cytochrome P-450 in, 153, 156
yeast, assay of genotoxicity of chemicals in, 10
yeast assays, constructed by recombinant DNA methods, 195
yeast β-galactosidase measurement, to assess inducible DNA damage, 195
yeast, biochemical markers in 14

yeasts, biosynthesis of cytochromes P-450 in , 131–139
yeast, biotechnological applications of, 42
yeast budding, coincidence with DNA replication, 183
yeast cell cycle, 102–105
yeast, cell cycle and mutagenesis, 102–015
yeast, centrufugal elutriation of, 186
yeast centromere, 14
yeast chromatids (bivalents), 14
yeast chromatids, crossing over of, 14
yeast chromosomes, 174–178
yeast, *cis*- and *trans*- acting elements, 47
yeast clones, toxicity of chemicals in, 181
yeast colony type, and genetic change, 180–181
yeast, continuous culture, for cytochrome P-450 production, 135–136
yeast, CYC1 genes in, 59–60
yeast cytochromes, general discussion, 189–190
yeast cytochrome P-448, benzo(a)pyrlne hydroxylase activity of, 143–148
yeast cytochrome P-450, general comments on, 190
yeasts, cytochromes P-450 occurrence in, 130–139
yeast, D4 strain, 174
yeast (D4, D6 and JD1) in test systems containing endogenous cytochrome P-448/P-450, 189–191
yeast (D6) in detection of nonmutagenic species, 194
yeasts, D7 strain, 174–176
yeast, demethylation of lanosterol in, 139–143
yeast detection systems, and tumour promoters, 194
yeast, diploid, sporulation check on, 180, 181
yeast, DNA repair in, general considerations, 74–77
yeast, dominant genes, 14
yeast, dose level of test compounds, 180
yeast drug metabolism, effect of phenobarbitone, 28
yeast, drug resistance, 32–34
yeast, drug resistance in, 49
yeast, drug screening in, 12–40
yeast endogenous plasmid 2μ (circu 2+), 50
yeast enzyme expression level, use of lac Z fusion, 51–52
yeast enzyme induction, 56–61
yeast enzymes, in benzo(a)pyrene metabolism, 143–148
yeast, ergosterol biosynthesis pathway in, 139–143
yeast, erythromycin resistance in, 32–34
yeast, expression of mammalian cytochrome P-450, in 195
yeast, fluctuation tests, 182
yeast GAL genes, induction of in yeast, 57–58
yeast, gene as functional entity, 42, 43, 44
yeast genes, epistasis groups, 87, 89–92
yeast genes, for iso-cytochrome C production (CYC1), 59–60
yeast gene fusions, 47, 48
yeast gene fusions, UAS and TATA box, 47
yeast gene, GAL2, 174
yeast, gene induction, 41–72
yeast, gene induction by DNA damage in, 85–86
yeast gene introns, in actingene and MAT α1 gene, 46
yeast gene LEU1, 174
yeast genes, linkage of, 14
yeast genes, protein repressors of, 55–56
yeast gene recombination, 14
yeast, gene repression, 41–72
yeast genes, RNA2-RNA11, 47
yeast, gene, TRP1, 49
yeast gene, TRP1, use of in yeast mutants, 51–52
yeast gene, TRP5, 174
yeast genetics, 12–40
yeast, glucose repression in, 57–58
yeast, glutathione peroxidase, 27
yeast growth, and tetracyclines, 29–32
yeast growth, effect of chlorimipramine, 21
yeast haem biosynthesis, 59–60
yeasts, haploid form, 169
yeast, heat shock effects on protein biosynthesis, 66–68
yeast heteroallelic mutations definition of, 15–16
yeast, heterologous protein, expression of, 42
yeast, high copy number plasmids, use with 46, 47
yeast, hybrid promoters constructed in, 47, 48
yeast hybrid proteins, value of, 52
yeast, incorrect glycosylation of proteins in, 42
yeast, induction of acid phosphatase, 60–61
yeast, induction of cytochrome P-448 by benzo(a)pyrene in, 148–151
yeast, induction of cytochromes, P-450 in, by ethanol, 137–139
yeast, induction of GAL genes by galactose, 57–58
yeast, induction of MEL1 gene by galactose, 58
yeast introns, 44
yeast introns, removal (splicing) of in transcripts, 42
yeast introns, seven nucleotide sequence in, required for processing of, 46
yeast, lariat-introns in, 46
yeast LEU2 gene, as selectable marker to detect transformation by exogenous DNA, 48
yeast, life-cycle of, 12–17

yeast, maltase induction by maltose, 58–59
yeast maltose permease induction by maltose, 59
yeast MAT genes, 61–66
yeast, mating factor α, 12
yeast, mating factor a, 12
yeast mating illegitimate, 179
yeast mating type genes, induction of, 61–66
yeast meisois, 14
yeast messenger RNA, all have poly A tails, 44
yeast messenger RNA, leader sequences in, 44
yeast messenger RNA, no obvious Shine-Dalgarno sequence in, 44
yeast methodology, in tests, 10
yeast mitochondria, 22–40
yeast mitochondria, attack by carcinogens, 172
yeast mitochondria, pyrimidine dimer repair in, 81
yeast mitochondrial DNA digest, PAGE of, 36
yeast mitochondrial nuclear interactions, 35–38
yeast mitotic recombination, 17–20
yeast mutagen, ethyl methane-sulphonate, 180
yeast, mutagenesis in, general considerations, 95–106
yeast mutagenesis, threshold level for, 180
yeast mutants, conditional lethal (temperature-sensitive) forms, 47
yeast mutants, identified with magdala red, 14
yeast mutation analysis, using integrative vectors, 51
yeast mutation, spontaneous and induced reversion rate of, 180–181
yeast, N-demethylase activity in, 131
yeast, non-ability to splice mammalian DNA transcripts, 42
yeast, optimization of cytochrome P-450 in, 134
yeast petite colony formation by Adriamycin, 26–29
yeast petite colony formation, effect of carcinogens, 23–29
yeast PGK gene, ORF activator sequence of, 55
yeast, pink colonies of, 175
yeast, 2μ plasmid in, 49–51
YRp7 plasmid, in yeast, 49
yeast plasmids contain CEN (centromere), 50
yeast plasmids, replication origin (OR1), 49
yeast plasmids, stability of, 49, 50
yeast plasmids with Amp^r and Tet^r genes, 49
yeast, production of antibodies in, 42
yeast, production of cytochromes P-450, effects of oxygen and aeration, 136–138
yeast, production of HB_sAg in, 42
yeast, production of tissue plasminogen activator in, 42
yeast promoters, high expression, application for, 195
yeast promoter fusions, 47, 48
yeast protein factors, interaction with UAS, 55
yeast protein, over producton of, 50
yeast protoplasts (sphaeroplasts) use of, 48
yeast, prototrophic colonies of, 179
yeast, RAD mutants of, 78–85
yeast, recessive genes, 14
yeast recombination percentage, formula, 17
yeast, red (ADE_1) mutants, 18–20
yeasts, red pigment in ADE mutants, 170, 174, 176
yeast, repair-deficient strains, 172–173
drug resistance in yeast, 49
yeast resistance to tetracycline, inheritance of, 34
yeast (homoallelic), reverse mutation induction in, 181
yeast revertant colonies, problem of no increase in per plate, 181
yeast, ribosome assembly in, 47
yeast, RNA polymerase II of, 42–72
yeast, selective toxicity in non-revertants, 181
yeast, short poly A tails in messenger RNA, 44
yeast shuttle vectors, 49
yeast sphaeroplasts, plasmid uptake by, 48
yeast sphaeroplasts (protoplasts) use of, 48
yeast, splicing of transcripts and ribosome assembly in, 47
yeast, spontaneous mutation in, 97–99
yeast spores (disomic) in aneuploidy scoring, 187
yeast sporulation, conditions for D7, 186
yeast sporulation, good aeration in, 187
yeast sporulation medium (SPM), 14
yeast, spot tests with, 182
yeast, stationary-phase cells, problems in tests, 181
yeast strain D7, in mutagenicity testing, 174
yeast SUC2 gene, 57
UAS, TATA box gene fusions in yeast, 47
yeast test cultures, use of a haemocytometer, 183
yeast test systems, absolute and relative frequencies of positive result, analysis and interpretation of, 192–195
yeast test systems, ADE2 gene in, 178
yeast test systems, and cell-cycle, 186–188
yeast test systems, and diethylstilboestrol, 194
yeast test systems, construction of assays using recombinant DNA methods, 195
yeast test systems, control considerations, 193
yeast test systems, controls in, 192–195

yeast test systems, evaluation in a number of trails, 194
yeast test systems, evaluation of toxicity from, 192–195
yeast test systems, free radicals in, 26–29
yeast test systems, for antibiotics, 29–32
yeast test systems, for tetracyclines, 29–32
yeast test systems, genetic stability of, 180–181
yeast test systems, graphical method (yield of genetic events), 192
yeast test system, in liquid media (as opposed to agar), 182
yeast tests, interpretation and analysis of, 192–195
yeast test systems, lethal hits, 193
yeast test systems, measurement of lethal effect, 182
yeast test systems, meiotic (sporulation), 187
yeast test systems, methodology and use of S9-MIX, 189
yeast test systems, mitotic and meiotic recombination, 172–176
yeast test systems, overview of, 193–195
yeast test systems, plastic or glass bottle use in, 182
yeast test systems, post-treatment growth to remove arginine permease and alcohol dehydrogenase with, 185
yeast test systems, problem of binding of test compound to media components with, 183
yeast test systems, problems with exhaustion of media, 184
yeast test systems, protocols with, 181–188
yeast test systems, quantative and qualitative interpretation of data, 182
yeast test systems, solubility limits of chemicals, 180
yeast test systems, statistical analysis of data from, 192–195
yeast test systems, "treat-and-plate" method, 182
yeast test systems, treatment of logarithmic-phase cells, 183–184
yeast test systems, treatment of stationary-phase cells, 182–183
yeast test systems, time of plates incubation in, 183
yeast test systems, time of treatment with, 180
yeast test systems, use of endogenous cytochrome P-448/P-450 in, 189–191
yeast test cultures, use of Coulter counter, 183
yeast test systems, use of cyclophosphamide and benzo(a)pyrene as control chemicals, 180
yeast test systems, use of D7 yeast to detect non-mutagenic species, 194
yeast test systems, use of dimethyl sulphoxide ethanol or acetone as solvent in, 180
yeast test systems, use of MNNG, sodium nitrite, hydroxyurea and dimethylhydrazine with, 180
yeast test systems, use of sodium thiosulphate wash with alkylating agents, 182
yeast test systems, use of solvents, 180
yeast test systems, use of synchronized cultures, 186
yeast test systems, validation and hazard evaluation, 193–195
yeast test systems, validation of D4, D5, D7 and JD1 yeasts, 194
yeast test systems, viability assessment, 182
yeast test systems, viable mutants produced assessment of, 193
yeast test, two-sided, 192
yeast tests, 5% budding maximum with, 182
yeast tests, use of master cultures stored in liquid nitrogen, 181
yeast tetrad analysis, 17
yeast tetrad formation, 14
yeast, threshold level for mutagenesis in, 180
yeast, transcription *in vitro,* 41
yeast transformation, 48–49
yeast transformation, by exogeneous DNA, 47–49
yeast, Ty transposable element, use of, 51
yeasts, use of master-culture of test strains, 181
yeast validation, of test systems using D4, D5, D7 and JD1 yeasts, 194
yeast vectors, 49–51
yeast YRp7 plasmid, 49

zonal-rotor, for centrifugal separation of exponeutial phase yeast cells, 186
zygotes (diploid), in yeast, 12, 13
zygotes, polyploid, 178